THINKING OUTSIDE THE BOX

THINKING OUTSIDE THE BOX

*35 Years in Kitimat and Terrace,
Working for Alcan, Raising a Family,
and Living the Good Life*

Alan McGowan

Copyright © 2019 Alan McGowan

All rights reserved. No part of this publication may be reproduced, stored in a retrieval system, or transmitted in any form or by any means—electronic, mechanical, audio recording, or otherwise—without the written permission of the author.

ISBN: 978-0-9947512-1-8

Editor: Audrey McClellan
Cover and book design: Frances Hunter
Back cover text: Audrey McClellan

This book is dedicated to those many people who have been frustrated by others who will not think outside the box, even though the rewards of doing so are enormous.

AKNOWLEDGEMENTS

Thank you to the many people who encouraged me to continue writing after reading my first book of memoirs, *Riding in Style*. I was delighted and grateful to receive the emails and cards you sent, telling how you related to different stories in that book.

For their support in helping me write *Thinking Outside the Box*, I'd like to thank my editor, Audrey McClellan, and my book designer, Frances Hunter, for their patience, talent and expertise, and for everything they taught me working on the two books together.

And I'd like to thank my daughter, Sharon, for helping me put it all together and keeping me going on it, as well as my son Joe and his friend Louis Zabot, for their eye on the technical details in the stories.

Finally, I'd like to thank my partner, Mary Jane Hogg, for her boundless support, humour, companionship and love.

DISCLAIMER

Names have been changed in some instances.

CONTENTS

Foreword *9*

CHAPTER ONE   New Beginnings *11*

CHAPTER TWO   Instant Plant *35*

CHAPTER THREE   Family *61*

CHAPTER FOUR   Lessons in Politics and Theft *79*

CHAPTER FIVE   From Footloose to Financial Epiphany *105*

CHAPTER SIX   Major Project Coordination Planner *129*

CHAPTER SEVEN   The Early Riser Cooperative Bus Line *149*

CHAPTER EIGHT   Unusual Incidents and Encounters *165*

CHAPTER NINE   The Invention of the Dry Scrubber Process *183*

CHAPTER TEN   New Challenges *213*

Epilogue by Sharon McGowan *228*

Foreword

At their base, these are the memoirs of a young newlywed couple who left their hometown due to family religious differences and started a new life in an isolated "instant town" located 450 miles up the BC coast. I thoroughly enjoyed writing this book, and I hope these stories will help people by showing how you can make your life and your job a lot easier by learning to "think outside the box." Doing so can also save you or your workplace millions of dollars and improve lives. By thinking outside the box, I learned to put my hat in my hand and listen to the average worker, who knew a hell of a lot more about his job than I could ever realize.

CHAPTER ONE

New Beginnings

Kitimat, Here We Come *14*
Getting Started *16*
Our First House *22*
Sick Baby *28*
Which Is the Front Yard? *30*
Private Hardware Store *33*

Mary and Alan's wedding, 1954.

My wife, Mary, and I were married in 1954 under very trying conditions. Mary and her family were members of the pacifist Russian Doukhobor community in the BC Interior town of Castlegar. Her parents did not approve of her marrying "out" of that community, and to an "Englishman," no less. They disowned her and refused to accept our marriage, and we were forced to find a new life elsewhere.

On our honeymoon we drove to Vancouver. I had heard that Alcan, the Aluminum Company of Canada, was hiring men for a new aluminum smelter in Kitimat, so I applied and was hired as a painter. Unfortunately, Mary could not accompany me to my new home because there was no town yet and no accommodation for families—just bunkhouses for workers.

So I said goodbye to my new bride and got in a small floatplane to make the hazardous, twelve-hour trip up the coast of British Columbia to the new town of Kitimat, which, like the smelter, was still in the last phase of construction.

We were flying along, following the Inside Passage at an altitude just over 2,000 feet, when whammo! We were in a dense fog. The pilot nonchalantly dived down to 30 feet above the ocean, nearly smashing into an old freighter. With a quick flip of the wings he shot by it and landed on the water. I don't know how close we came to the side of the ship in the dense fog, but we didn't miss it by much.

The pilot acted like this was an everyday occurrence. He taxied the plane seven miles up a channel in the fog and docked at a tiny store and café, where his six passengers had a lunch of oysters fresh off the local rocks.

The fog slowly cleared, and a few hours later away we went, arriving in Kitimat, where we were met by a man from the Alcan personnel department. He escorted us by boat—surprise!—to a barge out in the bay. A large bunkhouse on this barge was to be my floating home for the next month.

Kitimat, Here We Come

Aluminum smelters need a tremendous amount of electricity to operate, and there were vast quantities of water in the Kitimat area, so Alcan was able to develop a monstrous integrated hydroelectric power system that was one of the largest in the world. Over 1,400 skilled workers were needed to build and run the power system, the smelter and the new town. Alcan had advertised around the world, promising steady employment and a chance for advancement. A recession was on, and the Second World War had ended just nine years earlier, so young people from across Canada and around the world, particularly from war-torn Europe, jumped on board. I had seen the ad in *Maclean's* magazine before I had signed up.

When I arrived in Kitimat, I found people of all shapes, sizes and backgrounds. We had Canadians who were just out of prison, wanting to start over with their families where nobody knew them. (This didn't work in most cases—somebody always seemed to be there to recognize them—but it was no big deal; we were all in this together.) We had prairie farmers' children—second sons, who weren't in line to inherit the farm. And we had East Coast fishermen who had never had it so good, but who missed their families.

We had skilled German tradesmen and many German war orphans who had never known their mother or father. Some had never lived in a normal home, with normal people. All their lives they had lived in orphanages and on the street, eating out of garbage cans and surviving by their wits.

We had Scandinavian resistance fighters who wanted to forget the war and start over with their families. Sometimes mistakes were made: the Scandinavians would end up working with the Germans, and tempers would flare. There were Dutch men who also held big grudges against the Germans, and every now and then these would rise to the surface.

There were upper-class educated Brits who had gone broke in England during the war and were trying to start over. And we

Kitimat townsite, 1956.

had Catholics from Ireland who had been openly discriminated against at home. Plus numerous dour Irish Republican Army men who'd had to leave the country but were ready to go back at a moment's notice to fight for their country's freedom from the United Kingdom.

We also had Kiwis and Aussies, and later many Portuguese workers and their families arrived from the Azores. Even later the company brought in a group of skilled, British-trained tradesmen from Hong Kong.

Across the bay from the townsite and plant was the Native fishing village of Kitimaat, occupied by the Haisla people who had been there for millennia, fishing and hunting. Now they were being forced to live with 3,000 construction workers and 1,400 aluminum smelter workers and their families from all over the world. Many of the Haisla men hired on as well. The company had promised them jobs in the smelter, though some were reluctant to take the jobs because of the cultural adjustments that were required. It's no wonder that when we were working side by side in the smelter, some of them would look at all us outsiders cockeyed.

A lot of the people who came to work hadn't a clue about working conditions in a smelter. This generated a catchphrase that summed up the situation: "One crew coming, one crew working, and one crew going."

The start-up turnover was unbelievable, creating a great deal of chaos that sometimes put an equally great deal of strain on supervisory staff. They had to shuffle men around to ensure they would get along. After all, we had just finished a war where some of these men had spent five or six years trying to kill each other.

There was one other big problem. My smart young bride, when she finally arrived, observed, "There's not a single gray-haired person here." I realized they had only hired people between the ages of twenty-one and thirty-six. Alcan had a policy of hiring young people for a young town. This was short-sighted. Young people meant young families. Many young women were pregnant or had very young children. Thousands of miles from their mothers, they needed advice on pregnancy, birth, breast-feeding, diaper rash and God knows what else. It didn't help that we didn't have phones to call home. The only phone was in the radiophone office, and anybody with a shortwave radio could listen to your conversation for entertainment. At the beginning we had no local radio or TV stations either, which didn't help.

But I'm getting ahead of myself.

Getting Started

The day after I arrived in Kitimat I took a jitney boat from our floating bunkhouse to shore along with the rest of the ten-man group I had come up with. We were escorted directly to the plant's Stores department. There we were outfitted with heavy steel-toed boots and thick, woollen, fireproof pants and sweaters. Apparently this was to protect us from getting burnt by splashing aluminum.

I was confused. I had hired on through the Painters Union in Vancouver and wasn't planning on painting where I would be splashed by molten metal. But I followed the group down to watch an aluminum smelting pot started up. You couldn't see more than six feet for the black smoke, and the heat was intense. Not somewhere I ever intended to work.

Fortunately, on the way down I had noticed a sign reading Maintenance Office. I left the group and headed there.

At the counter I asked to see the maintenance supervisor.

"I'm confused," I told him. "I am a painter out of Painters Union Number 138 Vancouver, but they have outfitted me to be a potline worker."

The supervisor told me firmly that everybody was going to be a pot man, working the potlines, until the plant construction was sorted out. I told him to book me a flight back to Vancouver. I wasn't going to work in the potlines, even temporarily.

I must have raised my voice because an older fellow came out of the biggest office and, with a French accent, asked what the problem was. I repeated my story and the two of them went back into the supervisor's office and shut the door. I could see that I'd run into a bureaucratic screw-up. After a lengthy confab they came out and told me to get my kit and go up the hill, away from the plant, to the Construction Paint Shop. There I should see the foreman, who would "put me to work painting."

The next morning I dutifully reported for work, but the Paint Shop was completely deserted. Eventually the foreman appeared. He was a white South African with a nervous, fidgety disposition—he couldn't stand still for a minute, always moving, as if his pants were on fire. He was temporarily handling all the new painters, carpenters and labourers, and he told me I was going to paint the interiors of the small temporary houses located on the hillside right beside the smelter.

I found this strange. Why would they paint over a hundred houses that were going to be torn down in a few months when the permanent town was built? I asked the foreman some technical questions about

Small prefab houses and bunkhouse, Kitimat, 1954.

the job and he just shook his head; he didn't have a clue about painting. So I got my tools and started working.

I later learned that I had been hired a few weeks too early. Alcan had a crew of independent contractors painting the plant, and they couldn't let me work there until the contractors left. They needed me to get lost for a few weeks until the Plant Maintenance Department could take over managing the Paint Shop. So they found a make-work project to keep me busy.

After everything quieted down and all the construction people left, these 135 temporary houses were put up for sale at $2,000 each. When they were bought, they were also removed from the Kitimat site. Those little houses ended up being moved to various towns and other locations around the north country. Fifty years later, I could still recognize them, even though most of them had been modernized. Their weird small windows were unmistakeable!

* * *

A month later all the construction painters were laid off, and at the tender age of twenty-four I was promoted to the position of gang leader of the Paint and Glazing Shop. I was supervisor to twelve men who were not painters or glaziers. I had my work cut out for me teaching them the basics of painting, but I hadn't a clue how to cut glass.

Luckily, in the first two weeks nobody needed any glass cut or replaced. Slowly, practising on my own, I learned some of the skills and tricks the hard way. Glass cut when it is very warm is more forgiving than glass that is cold. And you have to dip steel cutters in a light oil before attempting a cut because the tiny cutting wheels can jam amid the tiny particles of glass.

One day Peter, an electrical foreman, came in with a work order. He waved it in front of me with a big smile on his face and announced, "I've got an interesting job for you. We talked it over in the Electrical Shop and came to the conclusion that you are the perfect man for this."

He sounded like a used car salesman as he explained this job. "You know the kangaroo crane at the wharf? It sticks way up in the air and hangs out over the water, 145 feet up? The red airplane warning light at the top of the boom is out and has to be replaced."

I replied, "That's an electrician's job! Go get an electrician to replace it. Why did you bring it to me?"

Peter sheepishly replied, "All my guys are scared of that height, and then we heard that you used to do high-steel painting."

I agreed to do the job as it was no big thing to me and would only take a few minutes. Peter was relieved and we jumped into his truck. But then he told me that he also had a red glass cover for the light that had to be replaced. When I asked what had happened to the original light cover, he laughed nervously and said he didn't know. It just was broken. I began to get suspicious.

At the crane I asked for the boom to be lowered to a 45-degree angle over the water and climbed out to the end. When I got there I saw that both the bulb and the light cover were smashed. This wasn't an ordinary burned-out bulb situation. I replaced the bulb and the cover, then came back down.

On my return to the truck, I demanded that Peter tell me what was going on. He finally confessed. "Some crazy bastard blew the light out yesterday with a high-powered rifle after firing about twenty-five shots.

They figure he was shooting from one of the bunkhouses over a quarter of a mile away. All my men were too scared to climb up there in case they got shot."

Of course, I wasn't too happy that he was telling me this now, after I'd risked my neck.

Two days later I had a visit from Gerry, the Chief of Security. He was a tall, handsome, middle-aged fireman from Saskatoon who had come to Kitimat with a group of firemen from his hometown. Gerry came up to the Paint Shop with a work order to replace a window on the third floor in one of the large bunkhouses. Putting two and two together, I asked if this was the room where the sniper was shooting out the lights on the kangaroo crane.

Gerry hesitated but finally said, "Yes, he broke the glass so he could get his rifle out the sealed window."

When we got to the bunkhouse, there was a guy in the room, relaxing on the bed.

I asked Gerry, "Is this the guy that smashed the window?"

Gerry said, "No, this is his roommate. The guy with the rifle has been fired and locked up tight until the next boat leaves for Vancouver."

I didn't think we had a jail, but he couldn't go anywhere anyway as we had no way out of the town.

I asked the guy on the bed what had made his roommate do such a stupid bloody stunt. He replied, "He's always been a little weird. He mail-ordered that sniper rifle with telescopic sights and had a couple of drinks after taking it out of the box, fondling it like it was a woman. Then he began looking out the window for something to shoot at. He finally spotted the red light atop the kangaroo crane a quarter of a mile away, and nothing would do but he was going to shoot that light out, come hell or high water. I was lying here calmly reading when he suddenly smashed the glass out of the window and started shooting. I took off out of that room, running like hell, and didn't come back until the security had hauled him away."

I measured the window and the next day came back and replaced the glass. All I could think was maybe the poor guy was "Bunkhouse Happy."

There were several weird broken windows in those first few months. Shortly after the kangaroo crane incident, I got another call in the Paint Shop. Someone, in a fit of rage, had thrown a beer bottle through the window of one of the gambling shacks.

I got my tool kit, called a dispatch truck and went down to replace the glass. The two gambling shacks could get pretty violent at night, but I was doing this during the afternoon, so I thought it would be safe.

I found the broken window, installed the new pane and then went inside to trim the putty. There were six guys sitting around one of the large circular tables, playing cards. This was unusual for a working afternoon.

I was busily working with my back to the guys when suddenly someone loudly declared, "I lost $40,000 last night. One of you bastards is cheating. That's got to stop or somebody is going to die."

This was followed by a dead silence.

I turned around to see a young man with his hand on a 9mm Lugar pistol that was lying on the table in front of him. The five other guys were staring fixedly at him. I couldn't gather my tools together quickly enough, and I got the hell out of there.

A couple of weeks later I got a work order to paint the gambling shacks inside and out, so I went down to take a look at what was needed. I had heard that they were hiring a professional gambler out of Vancouver to supervise the gambling. What a brilliant idea: hire a thief to catch a thief!

I walked in and who was standing there, all smartly dressed, but Harry, a professional gambler who used to ride the bus with me in Vancouver. At the time, Harry had been working at an exclusive gambling club and I was working at a moving company warehouse. We had shared some very interesting stories on that bus. Now it turned out that Harry had been hired by Alcan to control the gambling shacks.

We were chatting about the odds of both of us ending up in Kitimat when who should walk in but Marianne, a girl I had gone to high school with in Vancouver! What a small world. The last time I had seen Marianne, she was a "working girl" based in the old Martin Hotel in Vancouver. In high school, she and her friend stood out due to their outlandish style of dressing, which included parading around in full-size fur coats.

This seemed too much of a coincidence. Harry explained that he had brought Marianne up after he realized there were more problems in the camp than just gambling. The construction company building the town had turned a blind eye to a flourishing prostitution business. Even though there was no way in or out by land, fish boats were easily sneaking girls into the camp. Given that there were thousands of single men with money trapped in a closed town, business was booming.

The only problem was that if the girls stayed too long, some lonely customer was bound to fall in love with one of them and want her only for himself. Apparently this had happened two weeks earlier. The guy pulled a knife and blood was spilled. So the security police had asked Harry for advice, and he suggested they bring in a Madam to manage the girls and the Johns. The company agreed, and Harry contacted Marianne. From then on, Marianne ensured that girls were rotated in shifts every two weeks so no John could get overly attached to any one girl.

Harry gave a chuckle and stage-whispered to me, with a twinkle in his eye, "Now don't call me a pimp, I don't need that. This falling in love with a prostitute is a well-known hazard in 'the trade,' and usually the Madam can spot it ahead of time and move the girl on."

Our First House

By the fall of 1954 I had managed to rent a room from another young couple with a house, and Mary was finally able to join me. Mike and Sophie, the young couple, agreed to put us up for five months until construction on our own house, a duplex, was finished.

When we moved in with Mike and Sophie, there were only twelve completed homes in the entire townsite. Contractors all over Kitimat were desperately trying to complete at least 1,400 houses for the smelter workers, who were pouring in by the hundreds each week. Several months later, with organized confusion, there were over 700 completed and occupied homes.

Those lucky enough to get these first 700 houses were inundated with requests from other workers wanting to rent rooms in order to bring their families up, so now all these beautiful new houses were jam-packed with too many people in not enough rooms. And the building boom was so chaotic, all sorts of materials were lying haphazardly in the streets. People would just help themselves. We would be sitting around at night after supper and look out the window to see a sheet of plywood or a random toilet going down the back alley. And so it went, every night. Nobody seemed to get caught and, strangely, no night watchman was ever hired.

* * *

When Mary arrived in October by boat, she brought along my 1951 Pontiac Silver Streak four-door sedan. It was hoisted up out of the bowels of the ship and placed on the dock.

Mary also brought along the great news that she was expecting.

At this point the townsite where we were staying with Mike and Sophie was a sea of mud from the fall rain and all the building going on. What a place to bring my new, very pregnant bride.

The first time we drove up to Mike and Sophie's house, we saw two children playing on the side of the street. They were dressed in snowsuits and happily sliding down a 15-foot pile of soggy topsoil, plastered with mud from head to toe. Mary watched them, and I watched her face, but she never said a thing.

Along about November I came home from work and Mary had a gleam in her eye. She announced that we were going to go to the movies at the Smelter Site Hall. She had asked Pat, our next-door neighbour, to come with us. Pat's husband was busy and couldn't come. By this time, Mary, a big woman to begin with, was very big with child, and so was Pat.

Mary manoeuvred herself into the front seat of the car, I helped Pat into the back seat, and we drove down to the Smelter Site Hall. There were four huge bunkhouses at the Smelter Site, with over 2,000 men living in them. The films were shown in a very large hall or gymnasium. With nothing else to do at night, more than 250 men were sitting in the hall, waiting for the movie to start, when we arrived. All the lights were still on full as I came slowly down the centre aisle with a very pregnant woman on each arm.

As we walked, all the talk in the hall stopped for at least two minutes. You could have heard a pin drop as I helped Mary and Pat off with their coats, exposing their large bellies. The ladies conducted themselves as if nothing unusual was going on, but the next day at work I took a ribbing from some of my painters who were at the show.

* * *

We walked all over the townsite the first night Mary arrived and on every weekend afterward, trying to figure out where we would like to live. All we could see at this point was clay, mud, dirt and gravel, plus house foundation holes.

A month later I got a letter from the company telling me that I was now eligible to buy a house. I was invited to go to the one and only local bank, where I could sign up for a loan of $700, which would be my deposit on a $14,000 house at 3 1/2 percent interest. I was also invited to come to the company's Townsite Shack to pick out the lot for our house and the design of house that we wanted.

Signing up for the houses reminded me of signing up for the army. Each night forty to fifty men lined up and waited, then entered a long shack divided into two long rooms. The first room had a large map hung on the wall. You pointed to a lot on the map where there might be a partially built house, or only a hole in the ground, and you put your signature on your chosen lot. You then gave your name and number and picked out your house design, either a single house or half of a duplex. A lawyer gave you the legal description of the property, and you went on to the next room, where you gave a broker and salesman a cheque for the down payment. With that you became a home owner, but you had to wait months for your house to be finished.

The broker and salesman had been added to this process rather suddenly. Just before I received my letter, Alcan discovered that the government would not allow the company to sell the houses directly to employees. A licensed broker had to handle the sales agreements. The company frantically looked around for a local licensed broker. By a stroke of good luck, or a coincidence, a broker from Prince Rupert was in Kitimat visiting some relatives. When asked if he wanted the job, he naturally said yes, because there was a great deal of money to be made.

Rumours ran in all directions about how much the broker was paid for each transaction. Some said he got 2 1/2 percent of the purchase price; others said he got the down payment. By my calculations, 2 1/2 percent of fifty sales per night at $14,000 each equals $17,500. And handling the sales of all 1,400 of the town's houses would have been $490,000. That's in 1954 dollars. Today it would be in the millions. Not bad wages for a month's work. It's nice to be in the right place at the right time.

As Mary and I tried to decide which lot we should choose, in which area of the new town, I spoke to a lot of people. One old carpenter who was building the houses told me that in some low-lying areas the houses were being built on clay, which would mean basements full of water with Kitimat's heavy rain. Everyone recommended the upper level of the townsite, which had a lot of gravel for drainage. This area also overlooked the ocean, so that's what we chose.

We finally moved into our new duplex in January. The furnace ran full tilt twenty-four hours per day, so I went up into the attic to check out the insulation and found it was just stacked in a pile in the centre of the attic. I installed all the pads myself, by hand, and then turned my attention to the windows. They were single-pane glass and were constantly iced up, with a half inch of ice on the aluminum frames. I covered them with plastic sheeting. Checking out the oil-fired furnace, I found the combustion chamber was plugged full of soft sticky soot, the result of an empty whiskey bottle inside the pipe that was plugging the chimney. I removed the bottle, and finally the furnace shut off once in a while.

Some of the single-level houses in Kitimat must have been designed for California, not the snowbound north coast. The outside lower walls of their dining rooms and kitchens, below the windows, were a single layer of half-inch plywood. There was no insulation and no interior walls, just one half-inch of wood separating the residents from the cold outside. Can you imagine that first winter's heating bills, and the number of frozen pipes?

The heavy snow would come off the cantilevered roofs of these single-level houses and pile up on the ground. The heat escaping from the house melted the snow on the eaves and turned it into giant icicles that enclosed the entire house. Can you imagine not being able to look out the windows? Not being able to let your children outside to play for fear they would be crushed by the huge icicles?

That winter they were taking women down south for their mental health by the trainload.

*　*　*

Most of the working-class people who came to Kitimat had never owned a house. Some had never even dreamed of owning one. But as I said above, if you had a job with the company, you could go to the Bank of Commerce and borrow the $700 required for the down payment, and the bank payments would be deducted from your paycheque over a period of time.

Our payments on a $14,000 home, at 3 1/2 percent interest, were $80 per month while I was earning $160 to $200 per month. After a year of this, the union complained to the company that the payments were far too much of a hardship for the workers. So the company generously gave us a monthly bonus of $40 per month, which was applied to the payments.

Down the road, the union complained that the workers had to pay income tax on the $40 bonus. The company came through again, with a secondary bonus of $20 per month. In other words, Alcan was now paying nearly 75 percent of our mortgages. You couldn't beat this deal,

40 feet of snow, February 1965.

even though the union still complained that we had to pay income tax on both the $40 and $20 bonuses. It is no wonder that many people called the company "Uncle Al."

* * *

Our next problem was furnishing our three-bedroom house. All we had was a bed, four kitchen chairs and table, a coffee table, hope chest, camp stove and some dishes. We made out a list of what we needed, including furniture for a self-contained suite in the basement that I was going to build, even though I hadn't a clue how to do plumbing or electrical work.

I went to the bank and borrowed $1,000. Along with $2,000 that Mary had brought to the marriage, I figured we had enough. Especially since Alcan was generous enough to cover the cost of shipping said furniture the 450 miles from Vancouver.

My old painting partner, Orton Pelton, had given me a good piece of advice about buying my first furniture. He said, "Alan, your wife will want everything nice and new but don't go down that road. Buy good used furniture. Within four years, with all the babies throwing up and peeing on it, you will have to go and buy new stuff again." I talked this over with Mary and she agreed. Later on, after the two babies had nearly wrecked the chesterfield set and carpet, we got all new furniture for the living room.

I flew to Vancouver on a big spending spree, staying with Mary's older brother, Pete. For three days I went to auctions and used furniture stores, as well as private homes. I bought three beds, two stoves, three dressers, two fridges and a Bendix washing machine, two kitchen table sets, two chesterfield couches and chairs, two large carpets and I don't know what else.

Everything was shipped to Pier C in Vancouver and consigned to Alcan. I flew home, the conquering hero, with a purebred black Labrador puppy on my lap. We already had picked up a stray cat from the bunkhouses, where cats ran wild. Within a few weeks we were a complete family, snug in our own home with a brand new baby boy, Joseph Alan McGowan.

Sick Baby

Mary delivered Joe on March 1, 1955, in the old decrepit beach hospital. A week later we proudly took him home, and a week after that we had a very angry, crying baby who had a high fever and who would not eat.

We consulted the only older woman we knew, who had two teenage children. She came over and examined Joe and noticed a red swelling on one side of his stomach. She thought maybe he had a rupture and advised us to get him back to the hospital as soon as possible.

At the hospital the doctor carefully examined Joe and explained what was going on. "In a baby boy, the testicles are formed in his belly, and either just before birth, or within weeks, the testicles work their way down into the scrotum. In your baby's case, one testicle is in the scrotum, but the other one has pushed itself through the stomach muscle lining and is just under the surface skin. When he eats or moves, the stomach muscles tighten, putting pressure on the testicle, causing pain. This is a very unusual situation and requires an operation to rectify."

For the operation, the doctor needed special tiny tools and anesthetic equipment that had to be shipped up from Vancouver. Joe had to be kept under observation in the hospital until they arrived.

We went back to an empty house, devastated, and anxiously waited. Mary felt like she had done something wrong and was blaming herself. Each night we went to the hospital to see our barely one-month-old son, who, if we held him, screamed bloody murder. All we could do was stand there and hold his little hand.

Finally the equipment came and they operated. A few days later we took Joe home, thinking that painful experience was behind us. Boy were we wrong. A week later Joe had another high fever, so we took him back to the hospital and phoned the doctor. After a long wait he arrived, examined Joe, did something we couldn't see and gave us some antibiotic cream.

He told us, "If he ever has another fever, drop everything and bring him in. They will notify me and I will come right away, day or night." This made us feel that we had a dedicated doctor, and we took Joe home.

A week later the same thing happened. We went back to the hospital with our screaming baby, and the doctor did the same routine. The next week Joe had a fever again, but this time I took a closer look at the incision, which was red and swollen with a piece of black woven thread sticking out of it.

When we went back to the hospital, another doctor, I'll call him Dr. Morgan, was on duty and started examining Joe. Then our regular doctor came rushing in and said, "This is my case, Dr. Morgan. I'll take over now."

Dr. Morgan was surprised and said, "Very well, but I want to observe."

Our doctor briskly said, "Haven't you got other duties?"

Dr. Morgan said, "No, and I insist on observing the procedure. Alan here says that this isn't the first time this has happened."

I could see sparks were flying between the two of them and went outside for a minute. On returning I overheard Dr. Morgan say to our doctor, "You must have been drunk when you did this operation. You sewed this baby's stomach lining with exterior thread and his outer skin with interior surgical gut. You're a bloody drunk and should have your licence revoked."

I couldn't believe what I had just heard and got blazing mad. I grabbed our doctor by the shirt and threw him up against the wall. "If my baby dies, you die," I told him. "You're a bloody drunk."

Dr. Morgan pulled me off and said, "Alan, go into the waiting room and cool off. I'll be in charge from now on."

An hour later he came back and said, "We have removed the infected black stitches and cleaned up the wound, and you can take the baby home. Keep an eye on the wound, and if any more stitches come to the surface, bring him back in and we will remove them."

God, what a relief that we now knew what had gone wrong. Over the next month we took him in twice for the removal of more black stitches. I even pulled a couple out myself. Fortunately, Joe grew into a big, healthy, six-foot-three man.

Our doctor practised medicine for the next year and then left town. I guess we could have sued him, but being young and ignorant we didn't know how. We later heard that he had been drinking because he had just lost his wife in childbirth.

Which Is the Front Yard?

Our house was one of a long line of identical duplexes that ran along the main highway into Kitimat. The highway was on one side of the houses and the residential street (a dead-end cul-de-sac) was on the other side. This arrangement was normal, but one key thing was off. Every second house was turned around! The front door of the first house faced the highway, the front door of the second house faced the street and so on down the whole block. I learned that the general foreman had decided these duplexes would not look so monotonous if every second one was turned around. This went terribly against the master plan and had some interesting long-term consequences.

With baby Joe there was lots of washing to do. Because our front entrance faced the highway, I put a clothesline on the other side of the house, running from the back porch that was facing the street. This way Mary could come out the kitchen door and stand on the back porch, which was six feet in the air, to hang the washing.

One day when I was at work there was a knock on our back door. Mary answered it. Who should be there but a Bylaw Enforcement Officer.

This officious little man told my wife, in no uncertain terms, "The clothesline must be taken down because it is in the front yard." This despite the fact that he was standing on our back porch!

The idiot said that if we didn't take the clothesline down, the town would take it down and charge us for the labour.

The inspector was lucky Mary didn't throw him off the porch. It was never a good idea to demand that she do anything. I came home to a very agitated wife, and that evening I went to see the new mayor, who, luckily, I knew.

He just chuckled and said, "Obviously somebody is not thinking. I'll take care of it." I never heard another word about it.

This was not the end of the overzealous Bylaw Enforcement Officer. The next summer, in 1956, I worked very hard and built a beautiful white picket fence around the house. I had no sooner stood back to admire it than along he came to tell me the fence was 42 inches high and the bylaw said it couldn't be more than 36 inches high. He wanted me to cut six inches off the top of the whole bloody fence. I told him if he wanted six inches cut off the fence, he could cut it off himself.

Top: Petrel Street and our house (far right duplex).
Bottom: Sharon and the 42 inch fence.

That ended that problem, but having an issue like that hanging over us didn't make me feel good. We had enough problems getting along with everybody. We didn't need a Bylaw Enforcement Officer coming along, making up more trouble.

You give some people a little authority and it goes right to their head. A famous New York urban planner, Clarence Stein, had designed the town and given the municipality guidelines to follow, but he assumed it would use discretion to enforce those rules.

Private Hardware Store

As I mentioned, we had room for a self-contained suite in our basement. Now that I had all the furniture we needed for this suite, I started laying out a floor plan on scaled graph paper and listed the materials I was going to need to build it. My list included lights, a 220-volt outlet for the stove, a kitchen sink with all the plumbing, a toilet, a shower and a sewer drain.

Pricing these items at the newly opened builder's store in Kitimat, I found it was going to cost more than we could afford. When I mentioned this at work, one of the foremen dropped by the Paint Shop and told me to come to his friend's place that evening, as his friend had lots of extra supplies in his garage.

I walked over to the house after work. The guy was waiting with his garage door wide open, and I couldn't believe my eyes. He had almost everything I needed. But all these items still had the stock numbers from the Alcan supply store on them.

When I mentioned some items I needed that he didn't have, the guy said nonchalantly, "I'll have them for you if you come back in a couple of days." Believe it or not, he even had a stack of requisitions from the company store lying there, ready to be filled out.

I walked out, never to return. He was so casual about his operation, it was beyond me how he had not been found out. It was as if he had an inside source in the Alcan Stores department.

As the money came in over the next few months, I bought

everything I needed by dribs and drabs. But I still wonder how people can live the way the foreman's friend did. Unfortunately, theft was not unusual at Alcan, as you will see in a later chapter. And sometimes the thieves were very imaginative. But that's not surprising given the diversity of people who had come to Kitimat when the Alcan smelter got going.

CHAPTER TWO

Instant Plant

Dam Busters   *37*
You Didn't Shoot All of Them   *42*
Which Way Does the Wind Blow?   *44*
Corrosive Environment   *45*
To Prime or Not to Prime?
   (That Is the Question)   *48*
Stand Back   *50*
The Mystery Cheque   *51*
Eligible for Promotion, Just Not Yet   *53*
Gone Fishing   *55*
Locked In   *57*
Painting Cars   *58*

Alan at work, 1950s.

Take a random selection of people from all over the world, 1,400 men and their wives, plus their children, and put them in a super large box. Fly them up an isolated channel 40 miles from the ocean and 450 miles north of Vancouver.

This is a fairly good description of what we faced coming to Kitimat. Alcan bent over backward to help in any way it could, but, men being men, with their mix of ambition, jealousy, greed, ego, alcoholism, arrogance and God knows what else, people left in droves. It took over fifteen years to weed out the phonies, the shooting stars, the malcontents and the opportunists who make up a big chunk of the human species when social structure has not been established. Finally, trust in our fellow man was established and we could get down to operating a well-functioning plant that was pleasant to work in.

We had to have supervisors right away, however, so some big mistakes were made, with Alcan promoting people who didn't really know how to supervise. Many were put in positions way above their ability. We had staff members who didn't have the knowledge or experience and bluffed their way into staff positions. After a while they became scared to make decisions that maybe, down the road, might turn out to be the wrong one and could come back to bite them in the rear. So they were reluctant to make decisions that were desperately needed. When it came to the long grind of responsibility, a lot of men broke down and quit or were fired.

Some men had other reasons for leaving.

Dam Busters

The new boomtown of Kitimat had attracted people from all over the world, and nearly everyone had some kind of story to tell.

One morning in 1955 I came down to the wharf around coffee break and found four of my painters gathered with the machine operators and millwright in the long narrow unloader's control room overlooking the ocean. This was the only place with any heat on a cold rainy day, so everybody was sitting on whatever they could find, drinking coffee and eating sandwiches. Not wanting to interfere with their break, or the highly animated discussion, I stood to one side and listened.

We had a multiplicity of nationalities thrown together at Alcan, and surprisingly, in the early years, most of them got along well with each other. Actually, in those early days, everybody—and I mean everybody—went out of their way to get along. Many of these men were veterans of the Second World War, which had ended just ten years earlier. At this coffee break, I knew that Kurt, one of my painters sitting there, had served in the German army. And Dieter, an unloader operator, had served in the German air force. He worked with Charlie, who was British. You really did not know who you were talking to or what they might have been doing during the war, and you never knew when you were going to step on somebody's toes due to the war. So most men kept that part of their life to themselves. As time went by and we all got to know each other, we cautiously opened up a bit more.

On this morning, Jean-Paul, a very boisterous French-Canadian veteran, who had been a champion tree faller and was now one of my painters, was loudly holding forth on his army days. Suddenly he switched horses and lamented that he had always wanted to enlist in the air force, but because he had only a Grade Four education he was turned down. He looked kind of dreamy and said, "It sure would have been nice to be up there instead of down on the ground, in the mud and rain."

He pointed at Charlie, the crane operator, and said, "Charlie, weren't you in the British air force?"

Charlie looked at the whole crew and replied very cautiously, "Yes, I was." After a pause he spoke again. "It wasn't that great. More of us got killed than you army types."

Klaus, a Danish painter, spoke up. "Charlie, somebody told me that you flew fighter escort for the famous outfit that was called the Dam Busters. Is that true?"

Charlie slowly turned and stared at Klaus for the longest time. Then he looked around at everybody in the room and finally, quietly, gave a one-word answer. "Yes."

At this point our big-mouthed extrovert Jean-Paul, in a very loud voice, said, "What the hell are you guys talking about? What were these 'Damn Busters'?"

Charlie looked at him as if he had just crawled out of the cabbage patch, and started explaining what the "Damn Busters" were.

"During the war, the Allied Command got the bright idea that if they could break all the hydroelectric dams in the German Ruhr Valley, the devastation and loss of electricity to many German factories would bring the end of the war closer. The major problem with this theory was that you had to have pinpoint accuracy. You had to drop the bomb up against the inner wall of the dam. This was the only way you could break a dam."

At this point there were about ten questions thrown at poor Charlie. So he decided to tell everybody what dam busting was all about and how he had participated in it in the spring of 1943.

"Allied Command gave the problem of developing a strategy to get a bomb to explode underwater, against the inside wall of the dam, to an elite British air squadron. An English scientist came up with a radical theory. He said that they needed to develop a round bomb, and they had to figure out how to drop it from a plane just a few feet above the water, at a given speed and a certain exact distance from the dam. It would bounce across the water, slow down when it got close to the dam, then gently hit the dam, sink, and explode, using the detonation plus the water pressure to break the dam.

"Having one imponderable, is bad enough, but having five, was a tough nut to crack. They experimented with different bomb shapes, drop heights, speeds, distances and weights. After many tries they finally came up with a bomb shaped like a barrel, weighing 6,600 pounds, that they would drop from 60 feet above the water, exactly 420 yards from the dam, while flying at exactly 220 miles per hour."

Charlie said, "This was the point when I got involved as a fighter plane escort, and it was all very hush-hush and secret. I was transferred from my regular fighting squadron to the Dam-Buster Squadron, to provide escort for the bombers.

"Oh, and there was another thing. Just as I got there they were working on some form of machine that would spin the bomb at many hundreds of RPM inside the plane before dropping it, so it would travel smoothly across the top of the water. The bomber pilots told me that sometimes the spinning barrel would go off-centre while they were running it up to speed, and it would nearly shake the plane apart. Three tons of spinning bomb is nothing to play with."

Charlie stopped to get a breather, and mean old me stepped in and told my men, "This is not a working man's paradise. We have exterior painting to do."

Everybody got up and went to work, but Charlie's story was so interesting that I decided to stay around and eat my lunch with my crew in hopes that somebody would bring the subject up again.

Lunchtime came, and, sure as God made little green apples, before it was half eaten, Jean-Paul bellowed out, "Charlie, did you see any action?"

By now Charlie seemed a little more relaxed, though he responded, "I really don't like to talk about that."

Jean-Paul argued, "Luckily we are away up here in the godforsaken coast of British Columbia, in an isolated new industrial town. Nobody is going to mind. We would like to know how it all worked out."

Charlie replied, "You forget one thing. Sitting right here are men from around the world who were involved in the war, on one side or the other. Some can get pretty touchy when it comes to talking about the war. Thousands of people were drowned when those dams were broken and flooded the valleys below. How do you know that one of them here didn't have their family affected by those floods?"

Jean-Paul still insisted and finally persuaded him to continue his story, though Charlie only agreed to talk if Kurt and Dieter, the two

Germans sitting there, would agree. Both of them just slowly nodded their heads.

So Charlie began again, and the more he talked, the more relaxed he became. Maybe this was the first time he'd had a chance to let this all out. He began by describing the sequence of attacks before a bomb was dropped to bust a dam. "The fighter planes would make a first run by themselves to take out any machine guns or antiaircraft artillery on the dam."

As he spoke, I was sitting off to the side of the crew, out of the way. When Charlie mentioned a bombing run on a certain day in May 1943, I happened to notice a strange look come over Dieter's face. After a while he took a scrap of paper and a pencil stub from his pocket. He wrote something on the paper, then slowly put it and the pencil back in his pocket.

Charlie now seemed to be enjoying himself, with the whole crew listening to every word he said. He mentioned one unusual thing that happened on this particular run. "We had successfully made our first run, taking out the two machine guns, and had now looped around the back to escort the bomber in with his run.

"Just as we were approaching the bomb drop time, a Messerschmitt fighter plane came shooting up from below the dam with his underbelly right straight in my sights. I let him have both machine guns, and with smoke pouring out he went sideways over the side of the dam into a bunch of trees. That was my first kill.

"The Lancaster bomber let go of the bomb. It rolled across the surface and hit the dam, then sank and exploded but didn't break the dam. We all headed for home, and that night the squadron celebrated my first kill."

There was a long silence.

Finally Dieter quietly spoke up and asked Charlie, "You're sure that bombing run was on May 17?"

Charlie said, "Yeah, I should remember. I registered my first kill."

Dieter then asked, "Was that on the Sorpe Dam?"

Charlie looked at Dieter, puzzled, and said, "Yes, it was. How in hell did you know?"

Dieter then asked, "Do you happen to remember the numbers on the plane that you shot down?"

Charlie was now becoming quiet and cautious, but he said, "I'll never forget them. Sometimes I have nightmares about them."

Everybody now was dead silent, listening to this emotional exchange.

Dieter slowly reached into his breast pocket, pulled out the scrap of paper he had written on and showed it to Charlie.

"Christ! That's the same number."

Charlie had shot Dieter down and celebrated his "kill" with his squadron, and now they were working together thousands of miles from the Sorpe Dam, in Canada.

You could have heard a pin drop in that control room. They both stared at each other for the longest time.

Then, contrary to Hollywood bullshit, Charlie just said, "I'm glad you survived. Well, lunch is over, let's get back to work." Even Jean-Paul had enough sense to keep his mouth shut.

Charlie and Dieter worked a couple more years in Kitimat. Then both of them left separately for the United States.

You Didn't Shoot All of Them

Sometimes the encounters between former opponents were more explosive.

One day in 1962 I was at a meeting in the Forge Shop about a project I needed to have done when the door flew open and in charged Big Ernst, all six foot four inches and 240 pounds of loud, arrogant exuberance. Ernst could be a lot of fun if you didn't take him seriously, but he had recently been promoted to a foreman position, and the power had gone to his head.

Just a week earlier I had seen him blowing a whistle and lining up his crew of sixteen men to do calisthenics. A stranger sight the potlines had never seen. These men in hardhats, with heavy, burn-resistant boots and thick, woollen, fireproof pants and shirts, were doing strenuous physical movements in the smoke and heat after hours of heavy manual work among the smelting pots at temperatures up to 1,900°C. Needless to say, somebody in upper management put a stop to it quickly.

This day, Ernst was waving a work order and roughly pushed me away from the counter. I had been explaining my project to George, the gentle-mannered planner in the Forge Shop. Now Ernst demanded George's complete attention as he loudly exclaimed that pot rooms had first priority on everything.

Poor George looked at the work order and said, "Ernst, you want us to make a bookshelf, but you haven't included a drawing of what you want or indicated how big it should be. And this bookshelf has no production priority. You will have to wait three weeks, as the tinsmiths are very busy with high-priority jobs."

Stan, George's foreman, was sitting in an office next door, listening to all this. Stan was a quiet, middle-aged, English tradesman who had apprenticed as a tinsmith during the war, doing sheet-metal work on British fighter aircraft.

Ernst barged by me, straight into Stan's office, yelling, "The work for the potlines has priority over everything, and I demand to have this work order filled right away."

Stan calmly looked up and said, "Good morning, Ernst. What can we do for you?"

Ernst said, "I want this job done now."

Stan quietly replied, "It will be completed in three weeks, just like George has told you."

Ernst stood there, towering over Stan, his face and neck slowly turning bright red. Then he exploded, "During the war I was a Messerschmitt fighter pilot, and I shot down better men than you!"

There was a godawful silence.

Stan placed his pen carefully on his desk, rested his hands on the desk top, leaned back in his chair, looked up at Ernst and quietly said, "But you didn't shoot down all of them, now did you, Ernst?"

Ernst looked at Stan for thirty seconds, then walked out of his office. He plunked the work order on George's counter and slammed the door as he left, with not another word.

Surprisingly, Ernst made it all the way to retirement, many years later, without losing his zest for life.

Note: At the start of the Second World War in 1939, the German air force had nearly total control of the European skies. By 1944, however, Great Britain and the Allies had taken it back.

Which Way Does the Wind Blow?

In that first year I spent a lot of my time teaching my sixteen-man crew how to paint. Of the sixteen, four had some house-painting experiences, but steel painting in a large industrial plant was a whole different world and required close supervision for safety, particularly when it came to high-steel painting.

I had four men down at the wharf, painting all the steel outside and inside the two giant Symons unloaders—monstrous vacuum cleaners, 60 feet high and 40 feet square, that sucked the dry, powdered alumina from the holds of the bulk carrier ships onto a conveyor belt that moved tons of the stuff every hour to nine huge circular storage bins. The men were also painting the 145-foot-high kangaroo crane that was used to take various other products out of ships' holds using a two-yard clamshell bucket. We were trying to get everything painted before the fall rains set in, making the job more difficult. Salty sea air did not help the process. I also had teams of men spread around the plant painting offices, toilets and other equipment.

With four painters working full-time around the wharf, I had gotten to know "Old Mark," the wharf superintendent, pretty well. He could be as cantankerous as hell, though it was all a bluff, but he did rule his domain with an iron hand.

One morning I asked him to get his people to move their cars away from the north side of the office building to the west side. My painters were going to be spray-painting some machines there. If Mark's men didn't move their cars, I was worried the south wind would carry the aluminum paint overspray onto their cars.

With my "semi-order," Mark's hackles went up and he aggressively replied, "Listen, you young whippersnapper. I am an old sailor, and I've been predicting weather since before you were a spark in your father's eye. Don't you dare to tell me about our wind. Every morning at this time of the year at 10 a.m. the wind will change from the south to the north. We don't have to bother moving our cars."

Well, he came on so strong I thought maybe the old goat was right. So I left the boys with instructions to fire up the spray guns.

At 2:30 I went down to the wharf and checked the cars. Sure enough, they were all plastered with aluminum paint, and I mean plastered.

I braved the Beast's Den and went into Mark's office. He wouldn't admit that he was wrong, but he got all of his staff to go out and wash the paint off their cars with Varsol (a paint thinner). From then on he had more respect for me and my weather predictions.

Corrosive Environment

The construction company building the town did not want to waste time and money repairing tools when they got banged up or needed some maintenance. They just threw them out and got new ones, creating what locals called "the Million-Dollar Dump," which was located between the plant and town. All my life I have been attracted to garbage dumps, so I went to check it out. Security was supposed to discourage people taking things from the dump, but with no fences, and no hardware stores in town, it was open season. The dump was full of slightly damaged tools of all kinds that, with a little work, could fill every need. Sixty years later I still have the heavy-duty aluminum wheelbarrow that I found and fixed up with a new wheel.

On my first visit I noticed that the dump was packed with seagulls scrounging food dumped by the plant cafeteria. These were the dirtiest seagulls I had ever seen. The reason for that soon became clear. It was a cold, frosty fall, and I saw hundreds of gulls flying in large, lazy circles above the smelter complex, enjoying the heated air from the fans venting the 1,200-foot potlines where the aluminum was being smelted. Over the next few weeks the number of birds flying above the vents shrank steadily. I guess they realized the smoke was corrosive, and they were smart enough to avoid the fumes.

I soon discovered that this corrosive smoke caused problems all over the plant.

My Paint and Glazing Shop had just been moved to the new Building Trades Complex, inside the plant, when the phone rang and a voice at the other end asked, "Could you cut a piece of 20-by-24-inch glass and install it in the upstairs Instrument Shop in Building 157? The window just got broken." Then I heard a loud smashing of glass and the phone went dead.

I cut the glass and headed over. At the top of the long stairway up to the Instrument Shop I met the foreman, a short, stubby, middle-aged man with a twinkle in his eye. I asked him if he was the fellow who ordered a window to be replaced before it was broken, and he replied with a grin, "Yes sirree, that's me."

Laughing, I asked him, "Could you kindly explain what happened before I call security?"

The foreman picked up a piece of the broken glass and asked, "Can you see through that piece of glass?"

I replied, "No." The glass was heavily etched and impossible to see through.

The foreman said, "Neither can I, and it is my job to monitor the readings on the temperature and barometer gauges located right outside this window every day. The glass keeps frosting up due to the large amount of hydrofluoric acid in the air, and I can't see the gauges. Every three months I have to have the glass replaced."

When the plants' potlines smelted the aluminum, they emitted vast quantities of fluoride gas into the outside air through the hundreds of large roof fans. The fluoride blended with the humidity, or

Alan, 1950s.

rain, outside to form a highly corrosive solution of 5 percent hydrofluoric acid. This was what both the Instrument Shop foreman and the seagulls had learned. And there were many more problems to come.

My crew painting high steel with safety ropes found that after only three weeks they could break the one-inch manila ropes that were hanging outside with their bare hands. We could have used nylon ropes, but that was against Workmen's Compensation Board rules at the time, so we had to have special ropes made with polyester fibre.

The corrosive air was murder on the plant's 40-foot-long aluminum roofing sheets. The Cladder Crew, part of our Building Trades, ended up with a group of twenty-one men who did nothing but replace acid-eaten roof sheets, at a tremendous cost. As the gang leader of the Paint and Glazing Shop I suggested to management over and over during the next six years to test tar coatings or acid-resistant paint on the sheets, to no avail. Finally, after seven long years, the sheets were finally pre-coated at the factory with acid-resistant paint, extending their life tremendously.

The atmosphere was also hell on car windows, as we all painfully found out. If you brought your car inside the plant gate, you could expect to have all your windows frosted over within a year. Upper-level staff were issued permits that allowed them to park their cars inside the plant. This was supposed to be a special privilege, but the smart ones parked their cars outside the plant, away from the potline atmosphere and the prevailing winds.

All this changed twenty years later when one of our brilliant "thinking outside the box" chemical engineers discovered a process to capture the fluoride before it was released into the air. This discovery had worldwide environmental effects on aluminum smelters, dramatically reducing fluoride emissions to less than 4 percent of what they had been, but that's another chapter…

To Prime or Not to Prime? (That Is the Question)

One day in 1955, the Paint Shop received a work order from the Chief Residential Engineer. We were to paint the exterior of one of the plant's large Quonset huts with aluminum paint. This order came down through my Superintendent to my General Foreman, then down to the Brick Mason Foreman, my direct supervisor, and then to Little Old Me.

I took a walk out to inspect the oval-shaped building. It had served the construction people well for five years, but due to the plant's acidic atmosphere, the zinc coating on the corrugated steel sheeting that covered it was starting to get thin, and the iron sheeting underneath the steel was showing and starting to rust. The building was beside the road, where everybody could see it, and was starting to look shabby.

Being relatively new to the company, I was not familiar with the policy that encouraged the use of aluminum wherever possible in any work or purchases—i.e., aluminum paint, ladders, windows, siding, sheeting, electrical wiring, etc. The work order's instructions to coat the outside of a zinc-coated corrugated iron building with aluminum paint went completely against my technical knowledge as a journeyman painter.

Aluminum paint will not adhere properly to iron sheet covered with a baked-on zinc coating. And aluminum paint would deteriorate rapidly in the highly corrosive fluoride fumes coming out of the smelter. I concluded that I should first spray a coating of zinc phosphate etch primer on the building to provide a good key to hold the aluminum finish coat. I wrote out my recommendation and sent it up the corporate ladder to the originator. Back came the tersely written reply in bold capital letters: "PAINT AS PER WORK ORDER."

What should I do? I could paint it as per instructions, watch it disintegrate over the next six months and then get blamed for doing a lousy job. Or I could paint it the correct way and say nothing. Quite a quandary.

After a few days of stewing I decided to paint exactly half of the building as per instructions and the other half the correct way. I picked two men I knew would do the job as I instructed and promptly forgot about it.

One day the next year I was called into the foreman's office. There sat my boss, the Brick Mason Foreman, and our Superintendent, with faces of stone. They wanted to know why the Quonset hut was turning all rusty when it had been painted just a year earlier. The Chief Residential Engineer, or "God" in their hierarchy, wanted to know.

I had been keeping an eye on the building, and, sure enough, one side was all rusty. But the other side was still perfectly good. I cut to the chase and told them what I had done—the compromise I had made between following orders and doing a good job. I showed them the work order and asked, innocently, "Did I do something wrong?"

Of course I learned you don't buck the system and get away with It.

The next year the coveted job of Paint Corrosion Coordinator opened, and I was not even considered for the position. They gave it to an old-time staff employee (not a union member, as I was), who didn't know one end of a paint brush from the other. I had to work with him to develop corrosion-resistant test sample paints for the whole plant. I'll give him credit though. In a year or so we were seeing eye to eye, and he did a good job.

Stand Back

The electrical rectifier equipment at the smelter probably accounted for 25 percent of the total money invested in the whole project. With this in mind, I felt it had to be a showpiece, so I had two of my best painters working there for over two years, making it something to see. As I checked in on my men over that time, I got to know the rectifier staff pretty well.

Bill, the foreman, was a short, heavyset German with a large head. He was a perfectionist with a brilliant mind. He'd learned his trade in the German hydro-generating plant in the scenic town of Dresden. Like me, he had been a teenager during the war and not old enough to join up. But he sure could tell some tales about Germany during the war.

Bill and his family were survivors of the Dresden Firestorm. He remembered the Allied planes flying over the city, dropping leaflets that told people to get out of Dresden because they were going to "firebomb" it. And he remembered sitting on a ridge a mile out of town, watching thousands of incendiary bombs completely destroy the city.

One day I came into the rectifier building to check on my painters and ran into an extremely agitated Bill. One of the new main generators in the plant's Kemano power station was acting up—the power was fluctuating wildly, which was a major disaster—and upper management in the plant was not listening to Bill's solutions or letting him go to Kemano to fix the generator. Bill said he had encountered the same problem on two different generators in the Ruhr Valley in Germany and knew how to fix it, but no one believed he could do it.

This continued throughout the week, with the generator acting up and Bill getting increasingly upset. But the following Monday when I went to the building, I discovered Bill wasn't there. His foreman told me, "After two weeks of fooling around, all the big boys finally listened and they have flown Bill over to Kemano to fix the generator. The only problem is they have now made him mad, treating him like they did."

By Wednesday power began flowing consistently again. Word of Bill's success was all over the plant, so I went over to the rectifier building to congratulate him.

He walked toward me down the long pathway between the giant rectifiers with a jaunty stride. We went into his office. He shut the door, then turned to me with a smirk on his face and whispered, "I showed those smart alecks. I was blazing mad that we had lost all this time when I could have solved the problem in fifteen minutes a week earlier. When I got there, all the wise men from headquarters were standing around the generator looking very full of themselves. I told all of them to stand back at least 25 feet from the generator in case of a problem. I knew that some of them would want to learn what I was going to do, but I wasn't going to let them see.

"The generator was running at the correct speed but fluctuating in power delivery. Shielding what I was doing with my back, I opened up the top exciter cover, exposing the brushes. I took my screwdriver out and tightened all the springs to the brushes. I put the covers back on and the generator started purring like a kitten. The managers all wanted to know what I had done, so I told them it was a 'tradesman's secret' and I walked away."

Bill told me not to tell this to anybody, but that was 1955 and this is now 2019. It's been sixty-three years, and Bill is long gone to that great powerhouse in the sky. So I guess I can reveal his tradesman's secret now.

The Mystery Cheque

Everything was rolling nicely along; we had a house, car, furniture, baby and steady paycheque. I was part of the union as a shop steward and a gang leader paid by the hour.

One payday I was in line to get my bimonthly paycheque when the pay clerk remarked, "You must have worked extra hard. You have two cheques to sign for!"

I opened the two envelopes. There was my paycheque, for $95 minus deductions. But there was another cheque for $138 with no deductions. Now what the heck was this for?

I went to see our other payroll clerk, a well-educated German with a serious demeanour. Giving him the cheque I asked, "What is this for?"

He looked at it casually and said, "Cash it."

I told him I couldn't cash it without knowing what it was for. Maybe somebody made a mistake and would want it back.

I went to see my foreman, who said he would look into it. The next day he called me into his office and said, "I took that unexplained cheque to our general foreman. He did a quick investigation and told me to tell you not to worry about it, just cash it."

Again I replied that I wasn't going to cash it until I knew that it was kosher.

My foreman said, "Okay, if you want to stir the barrel, I'll make an appointment for you to see the superintendent."

The next day, off I went with my cheque to see the Big Cheese. He greeted me from behind his large desk, and explained, "The accounting people could not find where the cheque came from, but it is perfectly good, and I advise you to just go out and cash it."

I brought the cheque home to show my wife, Mary, who used to run her own store and had experience with accounting. She listened to my story, examined the cheque and said, "Alan, it looks like somebody has made a serious mistake and is trying to cover it up. You have two alternatives. Cash it and say nothing, or create a problem that will make you look stubborn."

I chose to be stubborn, and the next day I went to see my fishing friend Chuck, the Personnel Manager. Chuck was a tall, happy-go-lucky individual with an extroverted personality, and we got along great.

As usual, he was sitting with his feet up on the desk. When I came in they didn't come down, so I was looking at him between two large boots.

When I started telling Chuck my story, he became very serious. "Alan," he said, "I have heard about your problem and so has nearly the whole staff. I'm tired of hearing about your bloody cheque. I'm ordering you to get in my car, and we will go to the bank and cash that bloody cheque, and I don't want to hear another word about it. Can't you get it through your thick skull? Somebody has screwed up, and to straighten it out will cost far more than you cashing that cheque."

So I now had $138 in my bank account that I didn't earn. I can be very stubborn at times, when I think I'm right.

Eligible for Promotion, Just Not Yet

By the spring of 1955, management realized that the construction contractors had left but had not finished painting all of the smelter. They instructed me to estimate the number of painters and the amount of paint and equipment required to finish all the exterior painting during the short span of summer.

I found the size of the job still to be done overwhelming. I currently had only seven able-bodied men—six who could paint buildings and one sign painter—whom I had trained over the winter. But we would need an additional forty-one skilled painters, plus a large amount of paint and equipment, including electric chairs, swing stages, spray guns and sandblasting equipment in addition to the equipment we had inherited from the contractors. I prepared an eighteen-page proposal setting out how I planned to complete the painting over the summer, and, surprisingly, it was accepted without a query.

Word spread that I was looking for painters. One morning I found eight men plunked on my doorstep and was told to make painters out of them. Apparently the potlines were overstaffed and they had sent their extra men to me. I was skeptical about how these men had been picked, or been "volunteered," to go painting for the summer, and I decided to try to weed out the bad ones and the phonies.

"How many of you have painted before?" I asked them.

All eight put up their hands.

Then I said, "We're going to be doing a lot of high-steel work. How many of you are not scared of heights?"

Again they all raised their hands. I decided I'd have to give them the acid test to weed out the phonies.

I took them down to the kangaroo crane on the wharf and started climbing its ladder, saying, "Follow me."

One man stopped at 30 feet, one stopped at 80 feet and two stopped at 100 feet. We all came back down and I sent the four phonies back to the pot rooms.

Over the next week I tested forty more men and got another thirty-one who could take heights. I made my original six men into what we called "lead hands," each overseeing the work of up to seven inexperienced men.

With my own responsibility level on this project I should have been promoted to paint foreman, but I was still an hourly paid gang leader. I mentioned this to my foreman. He said he would see what he could do about it, but a few days later he came back to me and said, "I talked to the superintendent, and he said he couldn't do anything because at the moment the company has too many foremen they don't know what to do with."

I let the matter drop and tried to carry on in spite of the awkward situation.

A month later my foreman called me into his office. "I have some great news for you," he said. "The potlines have three spare foremen, and they're going to give them to us as painters for the summer. Isn't that great?"

This news completely floored me. Here I was, the only qualified journeyman painter on my team of forty-eight men. And now, as a mere gang leader, I was expected to instantly train, as painters on high steel, six foremen who outranked me. I should have been promoted to a general foreman role.

It reminded me of that old adage "This man is eligible for promotion, just not yet." I figured I must have pissed off somebody well up the chain of command to be treated this way, and I think I know who it was: he of the Quonset hut and "PAINT AS PER WORK ORDER."

Gone Fishing

Of the forty-eight men on my crew, eight were Indigenous men from the Haisla Nation village of Kitimaat, across the bay from the smelter. With no road access, they came to work by boat each day, regardless of the weather. These men were very quiet, and very observant. Sometimes I would catch them with amused smiles on their faces, usually because of something that was said or done by us newcomers. The Haisla were all solid, productive workers with absolutely no fear of heights. They ended up being the best high-steel painters in the bo'sun chair, swing stage or electric chair, while one of the newcomers would hold the steadying rope.

These men also knew their knots. One of them, Heber, taught me how to tie a running bowline. Heber was a powerfully built man in his thirties and had been a logger and fisherman all his life. Now he was one of my lead hands, and over the next five years we became very good friends.

About the middle of April I noticed something strange was going on among the Haisla crew members. A few of them would be standing together, talking in hushed voices. If they saw me approaching, they would immediately stop talking and look away. Some phrases I happened to catch were "Fifty cents is real good," "Northerners coming in," "Biggest catch ever" and "Hired on in Rupert."

I asked Heber what this was all about. At first he avoided answering, but eventually he said, with a chuckle, "Do you know anything about commercial fishing? Well, there's a great deal of money to be made if the fishing is good because you don't get paid by the hour, you share with the rest of the boat crew. This is the year of an especially big run from Alaska. The best we've seen in ten years. Do you blame the guys for being excited?"

I suddenly realized what was going on. Panicking, I said, "Wait a minute, back the truck up! These are steady employees of Alcan. They can't just quit and go fishing."

Heber, with another chuckle, replied, "Hold on to your hat because they are going to go, come hell or high water."

"Are you going too?" I asked him.

"No," he replied. "I've been down that road, and I like the steady paycheques I'm getting now."

Sure enough, a few days later I lost four of my great Haisla painters.

I got replacement workers, but they were not comfortable with heights, so they became safety-rope holders. We carried on over the summer and completed nearly half of the exterior painting. It still bothered me, though, that we had to use aluminum paint, because I knew it wouldn't withstand the acidic atmosphere and would be gone in two years. Such is bureaucracy.

In the fall I heard that after the fishing season was finished, the four Haisla men wanted to sign back on. Apparently when Alcan took over the valley it had agreed that any Haisla man who wanted could have a job in the smelter. The company agreed to hire these guys back this time, but never again.

Eventually Heber quit the company. I loaned him all my private painting equipment so he could start his own painting contract business, and he did very well with it. I missed him though. Throughout our time working together I had learned a lot from Heber about the Haisla culture and history. He was very knowledgeable and was also active in fighting for the rights of his people.

Fifteen years later I ran into him at the Terrace airport. He was all dressed up and was carrying a briefcase. I asked him what he was up to these days.

He laughed and said, "Can't you see? I'm an Indian in a three-piece suit, working with the federal government to establish our ancestral land rights, and I'm off to Ottawa."

We both laughed.

I understand he did a good job for the Haisla Nation, and I was very proud of him. He was a great friend, and in 1990 he came to my Alcan retirement party and presented me with a wooden carving, which I treasure.

Locked In

One day my foreman gave me a work ticket to paint the control offices for pot room lines 1 and 2, so I went over to the two-storey complex with my notepad to make an estimate of the time, paint and equipment we would need. I walked through the large conference room into an office at the back, and since no one was around I sat down at the desk to write out all the details of the project. I shut the door and was peacefully doing my estimating when I heard the plant Works Manager and his five superintendents come into the conference room.

I was trapped! By the time I figured out what was going on, they had already started their meeting. Heaven forbid that I would come out of the office and interrupt the six leading bosses in the plant. So I decided to sit there until their meeting was over and then sneak out.

I knew the Works Manager as the nicest man you could ever meet. Everyone in the plant liked him. But the man I heard chairing that meeting was something else again. He brooked no excuses. Over and over I heard him say, "Dammit! Get It Done. No Excuses!" I could hear every word through those thin walls.

When the meeting was over I had a totally different view of the Works Manager and those five superintendents.

It was only years later, at a management training class, that I learned the strategies involved in getting a plant up and running, and where this man's methods fit in. The first Works Manager needed to be rough and tough, accepting no excuses, in order to get things moving. This was definitely the person I heard in that meeting.

The second Works Manager needed to get things under control.

The third had to set up procedures, and the fourth had to cut costs with efficiency. We went through all those four stages with four different Works Managers over twenty years.

Painting Cars

In 1956, Kitimat was still a closed, company town, which made it awkward for a small-time contractor to come in and make any money. Consequently, since I had a subsidized house and a steady day job as the painters' gang leader, I was in a perfect position to take advantage of the situation.

Staff members with money were having large private homes built, but in the early years there was a lack of private painting contractors. There were private building contractors, and they offered me house-painting jobs at great rates, so I went into the painting contract business nights and weekends. I hired six of the best men on my Alcan crew to help out. I did this for three years, until there was no more big money.

Never one to sit around, I then built a shop on the side of my duplex, and the next thing I knew I was in the car-painting business, again on nights and weekends.

I'll never forget this one fellow who brought his car over to my place. He wanted an estimate to have his old gray car painted a nice bright metallic blue. Metallic car paint was just coming into vogue, but it was extremely touchy to spray-paint without making a complete mess of the job. If the air pressure dropped even two pounds while you were painting, the colour and distribution of the metallic flake would change, and you would end up with blotches of various shades all over the car.

Plus you had to have an air-driven paddle in the spray gun to ensure the metallic powder in the spray gun tank did not settle, which would result in the same blotched finish.

I quoted this man a price of $120, and he reacted violently, yelling at me, "You cheat, you cheat," in broken English.

I managed to get a few words in edgewise and told him, "Take it or leave it."

He quietened down and said, "What does that mean?"

I explained, "That's my firm price. I do not haggle." By now I didn't want to paint his car anyway, especially with his attitude.

He said, "I do paint job with vacuum cleaner, spray gun, save much money." Off he went, and good riddance.

Two weeks later he knocked on my door, on the verge of tears. "Meester Painter, you come look at my car, and you please paint."

This should be interesting, I thought. People believe painting a car is easy, but it takes a great deal of skill to get everything perfect, and it didn't sound like things had gone well for this fellow.

I went out to look at his car, and I couldn't believe the mess he had made. He hadn't sealed off all the glass and chrome, so there was paint splashed over everything. That's not too bad, because you can remove the paint with lots of lacquer thinner. The real mess was the paint drips all over the car. With every stroke of his spray gun he had developed a line of drips, so every six inches from the roof to the running board was a drip line. I had seen amateur messes before, but this one took the all-time record.

He started crying and said, "My wife, she no want to ride in my nice car. She says, get car painted right, and don't be so cheap. What can you do to help me?"

Feeling sorry for the predicament that he had gotten himself into, I said, "The price of $120 still stands, but you have to do all the work of repairing that damage."

He said, "You good man, that make me happy."

I quickly realized that he didn't understand just what I had said. So I carefully explained what he had to do, and I took him into my shop and demonstrated how to clean up the overspray with lacquer thinner. Then I demonstrated how to water-sand all the 150 feet of paint weeps. Finally I showed him how to mask off the windows and the chrome work.

I told him, "You do all this properly, and I will come over to your house when you have it all done to inspect it."

Suddenly his enthusiasm was back, and he drove home a happy

man. A week later he called me, and I went over to check that he had completed the repairs properly. To my surprise, it was done nearly perfectly.

He brought the car over to my shop and watched me paint it, remove the masking and do the cleanup. The car looked reasonably okay. He and his wife were happy, and he paid me, with much heartfelt thanks.

All in all, over the next nine years, I painted more than ninety-five cars and trucks. Coincidentally, many years later I developed kidney cancer. When I look back on the crude and crowded conditions I worked under, with no fans or masks, I know where my cancer came from.

Among the 1,400 workers and their wives who moved to Kitimat there were many people with experience at some business or trade. These people saw a glorious opportunity to fill a gap, and before you knew it there were small—and not so small—sideline businesses springing up. Builders, painters, auto mechanics, shoemakers, babysitters, barbers, chiropractors . . . all started up enterprises out of their new homes. Eventually some branched out into commercial businesses and became millionaires.

CHAPTER THREE

Family

My Feminist Wife *63*
Joining the Pension Plan *64*
Another Sick Baby *65*
The Mercedes Car *67*
Eight Bullet Holes *68*
A Good Doghouse *69*
Getting Out *71*
A Broken Car *73*
Don't Mess with Fathers *76*

Family photo in 1960.

Mary and I arrived in Kitimat in 1954, and within two years we had two children, a house and a dog. We were also part of a whole new community. Life was never dull in those first years.

My Feminist Wife

After just over a year in Kitimat, Mary was getting "Bunkhouse Happy." Before we married she had run her own grocery store for over nine years, and now she was stuck at home with only one-year old Joe for company. We had a "family conference," and as a result, in 1956, Mary got a job as a cashier at the newly opened Shop Easy grocery store in the Nechako Centre. She worked four hours a day, and Joe went to a neighbour's house for day care. Mary really perked up as she was again relating to a lot of people, and that made her happy.

One Saturday the snow was coming down in drifts. I decided to drive the short distance to Shop Easy to pick Mary up after work so she wouldn't walk home through the blizzard. Her shift was not quite finished when I arrived, so I stood to one side to wait.

Mary, with years of experience running her own store, handled the public very well. But as I learned the hard way, she had a lot of spirit and was a strong feminist. In those days very few people, including me, knew what those words meant. Watching her that day I got a firsthand lesson.

A newly arrived couple from Europe approached her till. The wife was pushing the grocery cart and the husband was behind her. The husband then roughly pushed his wife aside and said something to her in their original language in a loud, authoritative voice. The woman cringed as if the man was going to slap her or worse.

When the man reached for his wallet to pay for the groceries, Mary spoke up loudly with what was clearly a well-rehearsed speech: "Mr. X, you are in Canada now, and your wife has just as many rights as you. Including learning to speak English. And she is entitled to half the money you earn to buy groceries for the house. Why don't you give her some money so she can learn how to handle it?"

The expressions that replaced the husband's arrogant look were something to see. Surprise, then rage, then docility. Mary was a powerful woman, and at five foot eight inches and 180 pounds she wasn't going to back down from any man, big or small. The husband sheepishly handed the wallet to his wife and stormed out of the store.

I thought it was extremely humorous until Mary gave me a dirty look. She was boiling mad, and this was serious business.

Mary worked at Shop Easy for twelve years, and many times when we were out in public during those years, women would shyly come up to her and say hello. If I asked her who the woman was, out would come a tale of discrimination and of the struggles Mary had encountered trying to do something about it.

Joining the Pension Plan

After two years of work I received a letter from Alcan inviting me to join the company pension plan. I read it and put it aside.

Mary picked it up and got that strange look in her eyes that said, McGowan, you better tread lightly here.

She looked at me and quietly, but dangerously, said, "I think you should consider this."

I belligerently responded, "They want to take a big chunk out of my paycheque, and I only get it back after forty years. To me that doesn't make much sense."

She said, "Do you realize that after a few years they will match your donation, dollar for dollar? You cannot get an investment better than that."

I replied, "I'll think about it."

One year later I got another letter about the pension plan. This time Mary said, "Don't you care for me and the children? If you die, I'm left with nothing. A clause in that contract will provide me with at least a small pension or a severance package."

Finally the light went on, and I decided to join. I guess the people who ran the pension plan knew human nature, as they had a clause that entitled me to buy back the two years I had lost. Sixty-two years later, I have been getting that invested money back again and again, ad infinitum. Of all the investments and schemes I have been involved with over my life, the Alcan pension plan has been, by far, one of the best.

But I'm a slow learner. In 1961, when I was put on staff at Alcan with a monthly salary, I pulled the same stupid stunt as I had done with the pension fund. I received a letter inviting me to buy Alcan stock at a 20 percent discount with funds deducted from my salary, and I turned it down. At that point in my life it seemed that every time I turned around somebody or something was taking a bite out of my monthly salary for things like the pension, income tax, heat, light, property tax or scholarship funds.

A year later, after talking to two of my colleagues who were very sharp investors and who said I couldn't go wrong because I could always sell the shares, I changed my mind. And, again, I bought back the year I had lost. I kept buying the stocks, and after twenty years the company matched my investment dollar for dollar. You cannot get a return much better than that.

Another Sick Baby

Our baby girl, Sharon, was born during the very hot summer of 1956. At two months old she contracted hives and a fever. We had a sleepless night tending to the little tyke.

Thoughtlessly, being the man and the breadwinner, I went to work the next day, leaving Mary with the problem. I came home from work later to a very distraught female.

I hardly had time to get in the door before Mary screamed, "I can't take any more crying. She's all yours!"

Joe, one and a half years old, was sitting quietly on the carpet, looking very scared.

Mary put on her boots, coat and hat and said, "Your supper is in the oven. I am out of here." She slammed the door as she left.

Dumbfounded, I took a look and saw that the large, swollen hives covering Sharon's little body were coming to a head. She wasn't sucking her milk bottle and she had a raging fever. She would only stop screaming when she got too tired.

Foolishly, I thought it was too late to phone our doctor, and I didn't know if the hospital would accept or treat a baby with hives.

With the town made up almost entirely of young married couples, there were hardly any older women to give practical advice about babies and parenting, but Mary had found one mature angel, Chris Stump. She had five children and was a great help to all the young mothers.

Desperate, I phoned Chris, and she came over right away with some disinfectant cream. Chris examined Sharon, quietly looked at me and said, "Alan, what you have to do is going to cause her a lot of pain, but it has to be done. Each hive has to be cut open and squeezed empty, and then we need to put on this ointment. We will do this in the kitchen sink, in a warm flowing bath."

Rolling up my sleeves, I sharpened a paring knife, and Chris held Sharon in warm water as I cut and squeezed each of the hives with my baby screaming her head off. Then Chris gently dried her and put on the ointment. Amazingly, within half an hour Sharon's temperature went down and she was sleeping soundly.

Chris went home to her own children. Exhausted, both mentally and physically, I went to bed. I don't know when Mary came home, but she was up with me in the morning. Sadly, we never discussed this event.

All the young mothers had a hard time in Kitimat due to the isolation, and the long winters didn't help. Men were on shift work, trying to sleep during the day, and with so much snow the kids couldn't go outside to play. It was a terrible combination. During the winter of 1961–62 we had over 40 feet of snow, and young mothers left town by the carload.

I have a picture of Sharon and Joe with their hands touching the peak of our duplex roof as they are standing on over 18 feet of snow.

It's not a very impressive photo because I was too close to them when I took it, and you can't see the total perspective of a house nearly buried in snow... all you can see is snow and two kids.

The Mercedes Car

Living in Kitimat, we were surrounded by people from all over the world. However, most of our close neighbours were like us, Canadians, or Germans who had arrived in 1954 with the first large group of workers and their families.

By 1959 we had five years of steady paycheques in our bank accounts, and people were starting to upgrade their cars. The Germans with old Volkswagens were looking at more expensive luxury cars.

Gunther, living two doors down from us, bought himself a nice new Mercedes 180-D four-door sedan. At least once a week he was out there washing and waxing this lovely black car, as proud as can be. He used to sit on his front steps and just look at that symbol of German wealth and privilege. I guess if you were in Germany during or after the war, owning a new Mercedes car was really something else.

One nice spring day I decided to spray-paint our new picket fence and our clothesline pole. I fired up the compressor, checked the wind and went at it with fresh white paint. All finished, I went in for lunch.

A few minutes later I heard a God-awful racket outside. It sounded like a lynch mob.

Looking out from the porch, I saw a group of men and women surrounding the Mercedes. When they spotted me, they waved for me to come over, so I put my shoes back on and went over to see what the fuss was about.

Gunther met me and exclaimed loudly, "You put that bloody white paint all over my car! What are you going to do about it?"

It appeared that, as I was spray-painting, the wind had changed direction and carried the fine overspray down the block to his very special black car.

All Gunther's friends were speaking a mile a minute in German.

You'd think I had just blown up this fine example of German engineering and design.

It was clear to me that Gunther knew nothing about paint and car finishes. I went back home and grabbed a gallon of Varsol, as well as wax and some polishing cloths. I washed the white paint off with the Varsol, and, between us, Gunther and I applied a nice layer of wax. The car was good as new. In the meantime, seeing there weren't going to be any fisticuffs, his friends all went home.

Eight Bullet Holes

The houses on our street were built on a large gravel bar. When they were built, the company landscaper spread a very thin layer of fertilizer on the gravel along with lawn seed. It was a start, but nearly everybody had to haul in more of their own topsoil from out of the forest to have a decent lawn and to grow vegetables. We even went all the way to Terrace to get worms to aerate the soil, as there were no worms in Kitimat.

One hot summer day I was out trimming my new lawn when our neighbour, Stefan, who lived in the other side of our duplex with his wife and two daughters, came out with a push mower and started mowing his lawn too. Stefan was a heavyset, middle-aged German. We'd had a few talks about why he came to Canada and about his life as a soldier during and after the war. He was one German who swore he would never go back to the Old Country, but he had never given me details as to why he hated it so much.

Stefan got hot pushing that lawnmower, and he took off his shirt. For the first time I saw he had horrific scars from what looked like bullet holes all across his stomach and his back. Stefan noticed me staring and told me what had happened.

"I was a young man when the Second World War started," he said, "and I hated everything about Adolf Hitler and his policies. But I was conscripted into the army and ended up at the Russian Front. A Russian machine gun, at close range, shot those bullet holes in me, and my German compatriots left me on the battlefield to die. But I didn't die, and the Russians found me and took me to their hospital. They operated and saved my life. I was in that Russian hospital for over two years, recovering. I have nothing but good things to say about the Russian people compared to what I think of my so-called Fatherland."

It took me a while to say anything, and then I said, "I find it hard to believe you survived all those bullets, but I'm glad you did."

Stefan and his family were good neighbours for over fourteen years until he retired and they moved south.

A Good Doghouse

The word "Kitimat" means "Valley of the Snow" in the language of the Tsimshian people, the original residents of the area. And every winter we would get 10 to 20 feet of snow, although the actual accumulated snow on the ground by late winter would be only 6 to 12 feet. This was because we also got heavy rains that would melt the snow over the winter.

By February of our fourth winter in Kitimat we'd had the usual 10 to 20 feet of snowfall. One night, with about 5 feet on the ground, another six inches fell.

The next morning, Joe, nearly three years old, wanted to go out and play in the snow with Tragg, our dog. Mary bundled Joe up in his snowsuit and turned him loose in the backyard.

Mary and I were having a nice quiet cup of coffee when there was a god-awful roar. All the new snow came sliding down off our aluminum roof, falling two storeys into the yard.

Instant panic. Where was Joe?

I went flying down the back stairs, yelling, "Joe, Joe, where are you?"

There was no answer.

The snow off the roof was now piled up over 6 feet on either side of the house, and I couldn't find Joe or Tragg anywhere.

The company-built Johnson and Crooks duplex we lived in had tiny interlocking aluminum shingles on the roof. These were unusual, and excellent for the climate, but until that moment we hadn't known how dangerous this feature could be. Get four to five inches of snow on the roof and, whoosh, it all came off in a few seconds.

I yelled Joe's name again.

Finally a tiny voice said quietly, "Daddy."

I yelled, "Joe!" again and got another, "I'm here, Daddy."

I yelled again, "Joe, where are you?"

He replied, "With Tragg, Daddy."

Tragg had a large doghouse under the back stairs, and I realized that, luckily, Joe had crawled into the doghouse, following Tragg, just before the avalanche.

I grabbed a shovel and dug frantically down to the entrance of the doghouse. There they were, happy as clams, looking out at me with not a care in the world.

Sixty years later, in 2017, Joe and I went back and took a look at our old house. Amazingly, that aluminum shingle roof was still as good as new. Asphalt shingles would have been replaced at least twice in that time span.

As an aside, in the winter of 1961–1962 we had even more snow than usual. About 40 feet fell, but with the intermittent heavy rain it settled to a level of around 20 feet on the streets.

The town management had monstrous Sicard snow blowers that would, in less than five minutes, blow the snow off the road onto our yards and driveways. But in our so-called Planned Community, we were not allowed to push the snow from our driveways back out onto the road. Small personal snow blowers were not available like they are today. This meant that, day after day, you found yourself shovelling one to two feet of snow off your driveway and onto snowbanks that piled up on either side. By the end of winter these banks were more than ten feet high. Try shovelling snow to the top of a ten-foot bank two or three times a day.

By mid-February in 1962 I finally gave up shovelling, drove my car into a snowbank and left it there until spring. To hell with it; I took the local bus to work each day.

The town of Stewart, north of us, got 80 feet of snow that winter,

Eight feet of snow in 1958.

and the residents developed an ingenious strategy. When the snow piles reached ten feet high, the homeowners placed one-inch plywood sheets across the top of the snowbanks, creating long tunnels from the roads to their homes. No shovelling needed, except for the first few feet of the right-of-way.

Getting Out

By 1956 I had three weeks of holidays saved up, and boy were we excited to get out and visit relatives and friends in the "Outside World." At this point there was no road out of Kitimat. You put your car on a train to the nearest town, Terrace, which was 40 miles away, and then drove 1,000 miles of gravel road to get to the city of Vancouver.

The afternoon before you left you put on your old clothes and bought two bottles of whiskey, then went down to the Canadian

National Railway switching yard. You spent an hour chasing the railway crew around the yard to give them a bottle and ask them to please spot a flatcar at the ramp so you could load your car and tie it down with your own blocks and cables. Then you had a friend drive you home.

The next day, with suitcases, lunch, diapers, etc., you got another friend to drive you back to the station to board the passenger car of the train.

You arrived in Terrace a few hours later. After lugging your suitcases to a broken-down hotel and getting the family settled, you returned to the switching yard with the second bottle, flagged down the conductor, and asked him to spot your car next to the ramp. You'd put on your coveralls, take out your tools and a crowbar, remove the blocks and cables, and store them in the trunk of your car for the trip back home.

On Day Two you drove 350 miles on a very rough, narrow, gravel road to Prince George, the next big town, hoping that you didn't lose a muffler, have a flat tire, break a tie-rod or puncture your gas tank. (Over the five years of driving this road before it was improved, I experienced all of these setbacks.)

By Day Three of one of these trips, I was tired and irritated by the dust, rough roads and screaming, fighting kids in the back seat. When Joe spoke up for the fifth time that day and said, "Dad, I got to do a pee-pee," I grabbed an empty Coke bottle and thrust it back at him without braking. In a stern voice I said, "I am not stopping every fifteen minutes for a pee break. Now go and pee in the bottle."

Three-year-old Joe stood up, got his penis out and managed, on a rough gravel road, to push it into the bottle. He had just started peeing when I hit a large pothole. His little penis came out of the bottle and shot pee all over the back of my head and onto the dashboard.

Mary couldn't help herself. She started laughing. I stopped the car and we all had a good laugh. Lesson learned. I never did that again.

Later that afternoon, just outside Williams Lake, we finally got to a paved road and drove the 250 miles into Vancouver to start our holiday.

A Broken Car

By the fall of 1957, the 40 miles of road from Kitimat to Terrace had been pushed through and we took another trip to Vancouver. On the new road you still had to have a bulldozer pull your car through one area of hopeless mud swamp, and the 750 miles of gravel road between Terrace and Vancouver looked like a wrecking yard, as it was scattered with broken mufflers, shock absorbers and punctured tires.

We drove the first 600 miles in two days with a lot of heavy pushing, then stopped in Williams Lake to visit my mother and repair our car. The front shocks and springs on my prize 1951 Pontiac Silver Streak had been broken. I got the parts from a local agent and replaced them myself, while Mary and the kids visited Mom and her husband, Mitch, in their truck stop café, the Mixing Bowl. Then we had a nice vacation driving around British Columbia, visiting family.

As usual, the holiday was over once we started on that excruciating 900-mile drive home. In spite of bumping along on gravel for the first 600 miles, everything was going well until we hit a stream that suddenly appeared, crossing the road at the bottom of a hill. It was over two feet deep. The front wheels of the car sank into the stream, and the frame hit the road. We were only going 35 miles an hour, but the momentum carried the car up out of the stream and we coasted to a stop.

I got out to take a good look at the horrible sight. The front end of my prized possession had nearly separated from the body. The entire frame was broken just ahead of the firewall, and both sections were resting on the road.

To make matters worse, if that were possible, it started raining.

We hadn't passed a farm in ten miles, and I knew there was no habitation for the next few miles. Getting a wrecker in this desolate piece of country was a laugh. So I had to fix it myself. I got out our bumper jack and found a few pieces of wood, then jacked up both sides of the car and blocked them. I took my hatchet and cut down two hemlock trees, six inches at the butt and four inches at the top, 30 feet long, and ran them through the body, above the axles.

Leaving for holiday on the new road out of Kitimat, 1950s.

Taking a long walk around, I found an old abandoned section of barbwire fencing. Luckily I had all my tools with me, including a pair of fencing pliers. I cut a 100-foot piece of wire and wired up all four actual points, plus the bumpers. Surprisingly, I managed to do this without ripping my hands too much on the barbwire burrs. When I jacked up each side again and pulled out the blocks, she held together!

I spent over four hours doing all this work, and not one car had come along.

With all this separation, ripping and bending, I was worried that something might have broken on the drivetrain. I gently fired up the motor but nothing rattled, so I even more gently put her in gear and slowly let out the clutch. We were moving! I drove a few feet, then stopped, got out and checked everything.

The poor Pontiac. She sure looked weird, with those two long poles sticking out front and back.

Joe and puppy Tony.

Getting back in, I checked the steering. We were severely restricted due to the poles, with only about five degrees of wheel movement each way. We drove 30 miles very, very slowly and got to the small town of Burns Lake and its one garage just before quitting time. I parked the car and went into the office to talk to the owner, who was also the mechanic and welder. I will always remember the look on his face when he saw the contraption that I had made to get there. But he was too much of a gentleman to say anything.

We opened the hood, and he looked and muttered. I wondered what could be done, but I kept my mug shut. These backwoods mechanics had seen everything and had a world of experience in all directions, and I respected that.

The mechanic told us to come back the next day at noon and he'd have it ready for us. In the meantime, he gave us an old loaner car so we could get around.

We transferred our luggage and rented a cabin in town for the night. I suggested we go out for supper to celebrate, and everybody agreed. This seemed to take the tension out of the air. We went to a local restaurant, and as a special treat everybody ordered what they wanted. I was the only one to order liver and onions. I got food poisoning and was sick as a dog all night, and we had to stay another day.

Finally, after two days, I was well enough to drive, and I went up the hill to the garage. There sat my car looking as good as new! The mechanic had taken pieces of three-inch angle iron and welded the two breaks. The fix was far stronger than the original frame, and the welding was first class.

I braced myself and asked, "How much is this going to cost?"

He replied, "Thirty-five dollars."

In shock at how small the bill was, I exclaimed, "Thirty-five dollars?"

He looked at me, surprised, and said, "Is that too much?"

I said, "No, it's too little," and I paid him fifty dollars. That frame would never break again.

I picked up my family and we headed down the last 250 miles of gravel road, getting home without any more trouble. During the numerous times I passed through Burns Lake after that, I never crossed the threshold of that restaurant again.

Don't Mess with Fathers

Our house sat at the dead end of Petrel Street, next to a large cul-de-sac with a 20-foot circle of grass, a gravel area and a square parking lot. By 1960, every family on the street had one or two kids, so there were fifteen or sixteen kids of all ages playing in the cul-de-sac most of the time.

I decided to build some play equipment for all of these kids. In the centre of the cul-de-sac, without asking permission of the Bylaw Enforcement Officer, I put up a very stout pole and a basketball hoop. Off to the side of the grassy cul-de-sac, in the gravel area, I built a large "wobble board" out of railroad ties and car coil springs. In our yard, next to the cul-de-sac, I built a set of monkey bars and a swing set using pipes from the plant's scrap yard. In the winter I closed off the parking area on one side and put in a 15-by-40-foot skating rink, which everybody enjoyed.

A neighbour across the street took in boarders. One of these boarders was a middle-aged man who owned the first Mercury hardtop convertible in town. He was extremely proud of his car but was causing serious concern by speeding it around the cul-de-sac while all the kids were out playing.

Two of the mothers had spoken to him about his careless fast driving when the kids were around, but he just laughed it off. Mary mentioned this to me, but somehow I missed the importance of what she was saying.

A couple of days later, on a nice spring evening, I happened to look out the open kitchen door in time to see this fellow get into his fancy convertible and shoot down the street with screeching tires. He circled the cul-de-sac, barely on four wheels, and drove right through all the kids, including my son Joe and his friends. The car headed up the street and was gone.

I developed a cold rage, and when the driver came back I went over to see him. I told him that he had screeched around the loop for the last time.

He replied, "Ha, ha! And just what do you think you can do to stop me?"

Without raising my voice I told him, "You try that again and, believe me, you will find out. I am officially warning you to drive slow on our street, or you won't like the consequences."

He just laughed again.

The one virtue he had was that he was regular. I knew that tomorrow at the same time he would pull the same stupid stunt. So the next night I told the kids to stay away from the cul-de-sac. I took a gallon of used oil and poured it one-third of the way around the loop, right in line with the butt of a huge log, four feet in diameter, that was lying on the side of the loop with its butt end pointing toward the road. Then I went back in the house to wait.

I had the kitchen door wide open and was sitting at the table when the driver came out. He had the arrogance to wave at me as he got in his nice new convertible. He took off down the street as usual, screeching around the loop. Then he hit the oil and slid headfirst into the butt of that log, crashing his car badly.

With three other neighbour men I went out to survey the wreck. We approached the driver as he was looking at the car, antifreeze pouring out of the radiator.

"If there's anything you want to do about this, go ahead," I told him, "but, regardless, you won't be killing any of our kids."

He never said a word, and he didn't press charges, as I had a dozen witnesses to what he had been doing.

After everything calmed down and the car was towed away, I went out with a shovel and spread some sand and gravel over the oil so it wasn't so slippery.

Lesson: Don't ever threaten to harm any father's child or you could be in some very dangerous waters.

I recently read a book that, fifty-five years after these events, explained my violent reaction. When your babies are born they emit a pheromone, a chemical substance that you can't see or feel. However, if you hold and cuddle the baby, you breathe in those pheromones and are hooked for life, determined to do whatever it takes to keep that child safe. It's nature's way of protecting the young.

CHAPTER FOUR

Lessons in Politics and Theft

A Visit Home *82*
Mayor Alan *83*
Who's Stealing My Lumber? *85*
Never Expect Thanks *88*
Stealing *93*
  A free paint job *93*
  All wound up *95*
  Light-fingered Otto *96*
  Drop everything *97*
  Poor Owen *98*
  Salvage pass *99*
  Do you really know anybody? *100*
  WCB malingerers *102*
Shot on the Rise *103*

The Paint Shop, 1960s.

My unknown enemy must have quit or died, because in 1961 I was promoted to a position in a new field that the American Navy had invented called Maintenance Planning or Preventative Maintenance Planning. Alcan found out about it and got their training manual.

Basically, maintenance planning is a method of taking the guesswork and screw-ups out of breakdowns and maintenance work. To give people a sense of what I did all day, I used the example of an old steel bridge that is showing rust. The planner estimates the number of hours and the amount of paint and work needed to clean off the rust, prime the exposed steel and give it one protective coat of paint. Preventative planning additionally estimates the amount of time the protective coat of paint will last and schedules application of another coat of paint long before the rust begins to show. It's a great system, with long-term savings of labour and money.

Six of us were promoted to staff to be trained in this concept. After training, Alcan inserted us in various maintenance groups throughout the large plant. I became the Building Trades Maintenance Planner, with an office in the Building Trades Complex. I was expected to plan all the various jobs for the painters, carpenters, brick masons, cement finishers and painters—a total of forty men, with three foremen, three gang leaders and a general foreman.

In my new position I roamed all around the plant, estimating jobs and communicating with many different people. This was a staff position, which meant I was no longer in the union and I received a salary rather than an hourly wage.

I enjoyed the work, but it was a monstrous task to have thrown at me from out of nowhere. I worked my proverbial rear end off, trying to keep my head above water as I negotiated a steep learning curve, laying out work for all these various trades. The three foremen and the general foreman could see the load was too heavy, and they tried to help me out whenever they could. Finally, after three years, Joe, a tall Dutch technician, was hired on to help me by taking half the load. Then we had time to do in-depth planning, and things really started to roll along.

A Visit Home

A couple of years before my promotion, my boss, Scotty, who was foreman of the painters, bricklayers and cement finishers, decided it was time to go back to "the old sod" for a holiday. He had been in Canada for over ten years and was very industrious, and Canada had been good to him. He was proud of his accomplishments and it was clear he was planning to show off how well he was doing in "the colonies," with an important job, a new house and a new car. He had taken pictures of everything to show family and friends in Scotland.

Scotty had learned his trade working as an apprentice for the City of Glasgow under Mr. Burns, a cantankerous old foreman and brick mason who he described as "a great tradesman but a miserable boss." Scotty was eager to tell Mr. Burns about his affluent life in Canada and especially that he was himself now the bricklayer foreman in a large smelter, with thirty-eight men under him. With great enthusiasm, Scotty and his wife flew back home with their two children.

Four weeks later, Scotty returned to work, but his ebullience was gone. The story came out slowly.

When he had gone down to the Glasgow City brickyard to see Mr. Burns on the second week of his vacation, he had timed his visit to coincide with the ten o'clock kipper and coffee break. Walking through the old brickyard, he noticed that nothing had really changed, including the dilapidated tiny brick mason shack. The same old smoke was coming out of the chimney, which meant that Mr. Burns, his old foreman, must be in.

When Scotty opened the same door that he had opened thousands of times before, there was Mr. Burns, his back to Scotty, a coal shovel in his hand. He was holding the shovel inside the coal-fired heater, frying a kipper, just like always.

Without turning his head, his old boss yelled, "For God's sake, ye wee Nipper, come in, and shut the bloody door, ye wee runt. You're always causing a draft, and the kipper won't cook. Did you no learn anything over there in the colonies? But then you were never too smart."

Scotty sat down on an old sack-covered bench without saying a word and waited for Mr. Burns to finish cooking his kipper. When he finally turned and flopped the kipper onto a piece of board on a tiny table, he proceeded to eat it before turning to Scotty and saying, "If you want coffee, it's on the heater, just where it always is, and I suppose you haven't forgotten where the cups are."

Scotty got himself a cup of day-old black coffee and sat down without saying a single solitary word. Then, according to Scotty, "The old bastard accused me of going around Glasgow with my nose in the air, showing everybody that I was better than them. Can you believe it?"

I'm afraid I had to believe it, because that's what Scotty had said he was going to do when he got over there. But I didn't say anything.

Scotty said, "I apologized to him and tried to explain the differences between the two cultures, and he eventually seemed less miffed at me. Gone were all the boastful stories I was going to tell him. We shook hands, but I left feeling as small as I did in the first year of my apprenticeship."

Mayor Alan

When the Kitimat smelter started up, Alcan had formed a council, the AF of L, composed of eleven different trade unions and the pot room workers. The council was a disaster. Can you imagine twelve different groups of workers trying to reach consensus on what they wanted?

I had served on this council as the representative of the Painters Union and was on the committee that negotiated the first contract signed with Alcan. When I was promoted and joined the Alcan staff, I had to step down from all the union work. But a few years before that, the union asked me to run for mayor in the municipal election.

When the sitting mayor, or reeve as he was called in the early years, announced he wasn't standing for re-election, several candidates

Sam Lindsay joined the race after the article was printed.

announced their intention to run for the position, many of whom were in management positions at Alcan. The union brass wasn't happy with any of them, so they asked me to take part in the race. For a lark, I agreed. Then, just before nominations closed, Sam Lindsay, a very popular company personnel manager, put his name forward. I knew he would make a good mayor, but I decided to stay in the race anyway. It would be an interesting experience.

The union printed flyers that were passed out at the plant gate, and I had to make speeches. There was a large political rally in the high school gym with over 250 people attending.

When it was Sam's turn to address the rally, he spoke with a quiet, clear, calm voice that had the audience in the palm of his hand. The result was Sam soundly defeated his opponents. I managed to get 339 votes to Sam's 1,400 or so. One of the other candidates never spoke to me again. Maybe he thought I had split the vote and made him lose, but I don't think anyone was going to beat Sam.

We ended up with a great mayor. Sam would talk to everybody. He held the position from 1960 to 1970, when he, sadly, died in office.

Who's Stealing My Lumber?

As Building Trades Maintenance Planner, it was my responsibility to maintain the stock for the carpenters, painters, brick masons, cement crew and cladders. Once a week I would take an inventory list for each group, check the stock and reorder from Stores.

Stacked in front of the Carpenter Shop were piles of various sizes of lumber. I noticed that we were using more two by fours than I had charge account numbers for. I would bring the stock up to the maximum of 3,000 board feet. Then we would have no work for two by fours, but the stock mysteriously would be down to 2,000 board feet. No matter how much I ordered the next week, it would be down even more than the previous week.

This piqued my curiosity, and I brought it to the attention of the carpenter foreman and the general foreman. They said they would look into it.

The next thing I knew, "Supercop" from Security came to see me. He was a middle-aged man with a dry sense of humour and the damnedest ability to find things. After checking around, he told me that my two by fours "had gone up in smoke"; to stop it, I had to go see "somebody important" in potlines 3 to 5. He gave a chuckle and said, "The only thing I'll say is that from what they told me, they've been put to very good use. That's all I'll say. Thank god it's your problem and not mine, because it's a dandy."

The pot rooms had number one priority in the plant. Sometimes they abused this special status, and I realized that when I started asking why they were stealing our lumber rather than buying their own, I was going to have to be very diplomatic. Luckily, when I was just starting out, earning a man's wages for the first time in my life, an older man told me, "When you are handling people, you can slide farther on honey than you can on sand." He drummed this piece of homespun wisdom into me, and it has helped me over many tough spots through the years. Now, off I went to the upstairs office in the pot rooms with this advice in mind.

Les, the supervisor, saw me, and right out of the blue asked, "You want to know what we have been doing with your two by fours, don't you?"

He told me that he hadn't known about his men stealing my two by fours until Supercop came around looking for them.

His crew was responsible for a hot, horrible and dangerous job: they would use a long, heavy, steel rake to remove the gases that built up on the underside of the potline anodes. (A brief introduction to smelting: the anodes are positively charged electrodes that run an electrical current through the alumina in the pot, which is molten because it has been heated to 1,900 degrees Celsius.) When the gas builds up to a dangerous level, creating what is known as the "anode effect," a light would come on and the men would have to do their dirty work.

According to Les, one day a particularly lazy worker had used a wooden broom instead of the heavy steel rake, and within a minute the light indicating the presence of the dangerous gas went out. The worker didn't know what he had done, but he sure was pleased: no more raking anodes. He pulled the partly burned broom out, put out the flames and saved the remains of the broom. The next time he had an anode effect and the light came on, he did the same thing. Eureka! No more sweating.

He started using brooms all the time. He told the other pot men his secret, and there went the broom supply. Everybody started using brooms. A brighter star on the crew then reasoned that if the broom would put out the warning light, maybe adding a dry piece of wood would do the same thing. Another eureka! Any dry piece of wood worked to remove the dangerous gases.

"Now this is where you come in," Les explained to me. "You and your crew go home at 4:30 p.m., but our work is 24/7, so the guys would sneak down to your lumber pile after you left, get a few two by fours. Ergo, you lose your lumber, but the broom disappearance problem is solved."

Taking a full five minutes to contemplate this weird situation, I finally said, "Thank you very much for that great explanation. If you would give me a charge account number, we would be pleased to supply you with all the two by fours you need."

Les gave me a number and everything settled down for a while, with his crew still taking the lumber from in front of the Carpenter Shop.

One day I got a phone call from Grant, one of our chief buyers, asking, "What in the hell are you doing with all the two by fours? Are you building a house on the side?"

I told him I would be right up to explain the whole thing, and away I went.

After I explained the situation, Grant looked at me and said, "You mean they take perfectly good number-one-grade two by fours that have been planed on all four sides, costing $600 per thousand board feet, and burn them in the pot?"

I had to say yes.

He said, "So I am going to order thousands of board feet of sanded two by fours per month, and you are going to send them to the potlines just to burn them? Has the world gone crazy? Christ, we can get reject wood from the Terrace sawmill for $250 per thousand that will do the job just as well."

We ended up ordering 2 x 4 rejects, but then the pot men started to complain that the reject lumber had too much bark on it and wasn't strong enough to be pushed around in the molten aluminum. I relayed this information to Grant, and he eventually found a logger who was willing to supply us with eight-foot-long poles, five inches in diameter, that would do the trick for $200 per thousand board feet. We shipped in a few pallet loads and dried them by the pots, and everybody really liked working with them.

This led to changing the standard practice for putting out an anode effect throughout the plant. Everyone would use poles rather than grunt work and steel rakes. And everybody was very happy to eliminate a very unpleasant job.

This is a typical case of a so-called lazy man thinking outside the box and finding a much better way to do an important job. Let's just take a minute to analyze this scenario. As a pot man, you're instructed not to put anything in the pot because of the high chance of an explosion if there is any moisture present. Men have been known to cause a huge explosion of molten metal by throwing something as simple as an orange into an open aluminum smelter pot. So telling the foreman that you are solving an anode effect by shoving some form of wood into the pot might be construed as reckless behaviour. Therefore, the first man to do it didn't tell anybody until his idea was found out by mistake.

I'm sure that for the first couple of months, word of this simple solution for the anode effect spread slowly among the pot men. Did all the foremen see the piles of nice clean two by fours drying by a pot and wonder why they were there? Did they know the reason, but kept their mouths shut? Did the first so-called lazy pot man get any recognition for the solution he stumbled upon? No no no. Did some engineer take credit? I never could find out. Pots ran better, saving millions, and men didn't have to work so hard.

There are many more things invented this way than most people realize.

Never Expect Thanks

Shortly after I was promoted to my new position as maintenance planner in 1961, Nick, the general foreman, dropped in to my office and sat down heavily in the chair. This was very unusual. Nick always stood, in a very defensive way, said what he had to say and left. Something was up.

He started off by explaining that now that I was on Alcan staff, I was expected to take more of an interest in the development of the town, and to contribute somehow. He got me to agree with this idea, and then he sprang the trap. He told me that he had served two terms on the board of the local credit union. Due to other commitments, he was resigning, but he wanted to make sure there was someone who would replace him on the board. He thought that someone could be me.

I was a member and a strong supporter of the credit union, and I told him I would give this idea some serious thought. I'd have to decide if I could take this on along with all the other things I was doing. Besides, I said, "Elections to the board are not for another six months."

Nick looked kind of embarrassed and said, "Well, actually, I am resigning in midterm, and you could be installed tonight."

I hated to disappoint him, so I reluctantly agreed.

Our credit union was quite small, with capital of $120,000. It was upstairs in the service centre; had no full-time employees, just volunteers; and was only open Saturday afternoons. Tim, a big, heavy, happy-go-lucky Englishman who worked in the smelter, was the part-time manager and received a very small salary.

I went home, and after supper, with the two kids tucked away, I mentioned to Mary that I had been asked to become a director on the credit union board.

Mary said, "That's great! Now, finally, you will learn how to handle and save money." Because Mary had run her own store back in Castlegar, she was always frustrated by my lack of knowledge about money.

I jumped in the car, drove down to the service centre and climbed the stairs to the credit union office. Geoff, the president, greeted me and congratulated me on becoming a director.

I said, "Hold on! I am not a director until I'm elected and sworn in."

He looked embarrassed and said, "When directors resign, I am empowered to install a temporary director until our next general meeting, when you will be properly elected."

I noticed that Nick, who was resigning, wasn't there. Ian, another director and an acquaintance of mine, was also not there. There were only five of us, and Ian would have been the sixth, so I asked, "Where is Ian?"

Geoff flushed and replied, "Ian is not coming."

Wondering what was going on, I asked why.

Geoff, shamefaced, said, "Ian has also resigned, and we are looking for another pro tem director."

Until this moment I thought that I had been asked to sit on the board because I had something to offer. Now I was getting the impression that they were desperate for anybody. I smelled a rat and nearly walked out. Just what was I getting myself into?

Around this time a very distinguished-looking fellow in a suit and tie came up the stairs, quietly nodded to Geoff and went into one of the other offices. I recognized him as Jed, a professional accountant from Terrace. What was he doing here in Kitimat? The offices in the service centre were like fishbowls, with windows between them and into the hall outside, so you could see what everyone was doing. It looked like Jed, in the office next door, was going through accounting ledgers.

We started the meeting. Half an hour into the usual business of the credit union there was a knock on the door. It was Jed, with a sheaf of papers in his hand.

He said, "Gentlemen, I'm sorry to interrupt your meeting, but I have something very important to tell you."

Jed paused and then continued, "Mr. Chairman, I have just checked your financial records, and as of this day there is a total of $42,000 missing. I would strongly recommend suspension of all operations until the Superintendent of Credit Unions can get a team of auditor inspectors, up here to audit your financial records. What I have seen so far indicates some form of embezzlement. There could be criminal charges pending."

I began to feel like a new actor in an old play. This was some kind of planned audition, and I had not been given a script. All the other acting was fairly good, but I don't think anybody could take one hour to look at a messed-up set of books and then declare what he had with such surety.

Strangely, my first thoughts were not to save my own rear from the embarrassment of being involved with an embezzlement, but of the two rats who had known to abandon ship ahead of time so they could save their own rears.

What to do? I was in a perfect position to get up and walk away from the whole mess. Then I would not be tarnished like all the incumbent directors, president and manager who hadn't done their jobs properly.

I chose to stay on board and try to save our little credit union, although another rat quickly deserted the ship. Those three were partially guilty of creating the situation we were in. Why were they allowed to leave? It wasn't right or fair. (Many years later a law was passed that required directors of credit unions to sign a form making them financially responsible if there was money missing. This raised the quality and integrity of new directors.)

Our credit union was put under strict supervision, and the inspectors came in and thoroughly audited our books. They were in a terrible mess, and the inspectors found that the culprit was Tim, the manager. Slowly, the whole sickening story came out. Tim had married an English floozy who was out to bleed him dry. She wanted more and more or she would leave him. So she managed to get a fancy new convertible, jewellery, a fur coat, etc., thanks to our credit union funds.

We were under the supervision of the BC Central Credit Union for over a year. Plus, to allow us to keep going, the Central Credit Union gave us a loan of $60,000 that took years to pay back. I believe this was the first bailout in British Columbia credit union history.

Tim went to jail for two years. Personally, I felt sorry for him. He lost everything—his job with Alcan, his house, his wife, everything. And we had to pick up the pieces and carry on. Another rat director resigned. Thanks to Geoff, our president, I ended up as chairman of the Delinquent Loan Committee, on a very steep learning curve.

The committee consisted of two directors, and believe you me, we had our work cut out for us. We had to wade through thirty-five delinquent loans that hadn't been looked at in up to three years. Our job was to bring them all up to date and try to regain the missing money.

So, for example, say Mr. Jones countersigned a loan from the credit union for $2,500 so Mr. Billy could buy a car. Three years later we brought Mr. Jones in to be interviewed. A typical conversation went like this:

Chairman [me]: "Mr. Jones, we have in our records a granted loan application for Mr. Billy to buy a car worth $2,500. You co-signed the loan agreement. Mr. Billy has not paid one cent on the loan in the last three years, and we can't find him. Being the co-signer, you are now required, by law, to repay this overdue loan with interest."

Mr. Jones: "Who in the hell is Mr. Billy? I've never heard of him."

Chairman: "We have checked your signature against other documents and the signature Is yours."

Mr. Jones: "When was I supposed to have signed this note?"

Chairman: "Let us see . . . the date is May 3, 1958."

Mr. Jones: "What day of the week would that be?"

Chairman: "I believe it was a Saturday."

Mr. Jones: "Wait a minute, just a minute, something is coming back to me. I always try to go to the beer parlor on Saturday, and Mr. Billy was a workmate of mine. He came in while we were drinking and asked me to co-sign a car loan. He was heading back to Nova Scotia on a holiday and needed a car. Sure, I signed that paper, but I didn't think I was responsible if he didn't pay. Come to think of it, he never did come back to work."

Chairman: "I am sorry, Mr. Jones, but you are now responsible for this debt. To show good faith on your side, please sign this form that says you agree to pay thirty dollars per month until it is all repaid, with interest. If you do not sign, we will be forced to take you to court, and it will end up costing you a lot more than the $2,500."

Mr. Jones: "That dirty rotten bastard! I'm ending up paying for his car. If I ever see him again, I'll knock his block off."

Nearly every one of the thirty-five delinquent loans followed this scenario. Most of these co-signers worked for Alcan, and some of them I had to see nearly every day at work. This could be very awkward for me, particularly if I needed their cooperation on something.

I did this disagreeable task one night a week for over three years, as well as attending directors' meetings. At the end of the three years, at the annual meeting, with eighty members present, I stood and summed up what the Delinquent Loan Committee had accomplished in terms of money recovered.

Then I said, "I have served for three years, through the worst of times. I am tired and I am resigning."

You could have heard a pin drop. Several of the co-signers were in the audience. Not a thank you, no clapping, just empty air.

Lesson learned; don't expect to be thanked when you have completed a tough, painful and embarrassing job. Satisfaction is within yourself. Nobody can take that away from you.

Stealing

Shenanigans at the Alcan plant made sure I kept my eyes open. I was always astonished at what people would try to get away with, in spite of the serious consequences if they were caught.

I remember one day, early in my career, I was working with Bill, the rectifier foreman, a short, heavyset man who was extremely meticulous about everything. He was using a small fine file to clean up some silver contact points. When he was done he threw the file in the garbage can.

At this point I was still struggling to make ends meet. To me, that little file was worth a lot of money, and it looked like it had hardly been used. I picked it up and asked if I could have it.

He asked, "What are you going to do with it?"

I replied, "Take it home! I can sure use it, and, after all, you threw it away."

Bill looked at me very seriously and said, "If you don't take it to Security to get a pass for taking it, think of the consequences. You will be fired if you are caught, all for a four-dollar-and-fifty-cent file, when you are making $4,000 per year."

I never forgot those words and always used them to evaluate a temptation. Everybody has a different process for evaluating the pros and cons of taking something that doesn't belong to them, as the following stories will show. (Note: Names have been changed to protect the guilty.)

A free paint job

In my early years, when I was still painting gang leader, we were spending a lot of my painting budget down at the wharf, so I ended up getting to know a lot of the wharf personnel. Of all the characters, the one who stood out most was Mark, the supervisor, who had worked on Alcan wharves all over the world. He knew his job and ran a very tight ship. And he was as tight as a shark's ass when it came to his budget.

The offices that Mark and his three foremen occupied were in a dilapidated construction shack. The shack had never been painted, and the walls were nearly black from cigarette smoke and the residue from the bulk pitch and coke shipments that went through the wharf.

One day I caught one of my men painting the inside of Mark's private office, applying a nice, light green paint that was supposed to be for the wharf's control room. Mark had talked the guy into giving him a paint job paid for by my budget.

I put a stop to the work and confronted Mark. I said that if he would make out a work order to charge the cost of the paint job to his own account, I would forget about the attempted theft and let the guy finish the job.

Mark went red in the face and said he would get his clerk to take care of it.

Yeah, and the cheque is in the mail.

I waited for the work order but, of course, it never came. When I asked the wharf clerk where it was, he said, "Are you kidding me? There's no way in hell Mark's going to pay to have that old shack painted."

I decided to go to my next plan. I waited until the barometer showed that we were going to have a couple of days of heavy rain and then dropped into Mark's half-painted office. I told him that since we were expecting rain, my guys could move indoors and throw a coat of paint on the offices in his shack. He just needed his guys to move out for the day while we did it.

Mark brightened right up and said, "Certainly. There's no ship in, and there's no work anyway."

I told him to take everything off the walls and to cover any equipment they didn't want painted.

When Mark and his foremen came back the day after we finished, the ceilings, walls, floors and doors were all spray-painted with screaming bright aluminum paint.

Mark looked at me like a dog that had just been kicked and said, "I thought you would be using that nice green paint."

With a straight face I gave my prepared answer: "Well, Mark, consider yourself lucky. You got a free paint job. Because I didn't have a work order or charge account, I used what was available for a free paint job. It looks pretty good, don't you think?"

All wound up

One day, just as I was going through the Potline Maintenance Shop, a movement caught my eye and I stopped to watch two welders, Mac and Tosh, hard at work winding a new 100-foot copper welding cable around Mac's body. Tosh was holding the cable, slowly letting it slide through his hands as Mac turned, the cable winding around him up to his armpits. He then put his coat on over the 100 pounds of copper cable. A few minutes later I saw them heading for the security gate to leave the plant.

Now let's take a close look at this situation. They had 100 feet of new welding cable, weighing nearly 100 pounds. That is a fairly heavy weight to carry the 300 yards to the security gate. To wind the cable tightly around Mac's body so it would stay in place required a certain amount of pull or tension. Let's say 15 pounds of pressure each time it went around his body, which meant maybe another 100 pounds of pressure along with the cable weight of 100 pounds.

As he approached the gate, Mac started to stagger.

Then he collapsed right in front of the security officer.

The officer panicked. "Call the ambulance!" he yelled. "This guy's having a heart attack."

By now I was abreast of them and said, "He's just passed out because he can't breathe. Open his jacket and you'll know why."

Mac was fired. What are you going to do when you lose your job, and house, in a company town? Was it really worth the risk?

Light-fingered Otto

When I was the painter gang leader, one of my responsibilities was to look after all the equipment, including brushes, rollers, spray guns and various other tools.

A young fellow from the pot room had asked for a transfer to the Paint Shop because he had done some painting in Germany. He was nineteen years old but looked about fifteen, very slim, with bright, light blue eyes and curly, nearly white hair. Soaking wet he might have weighed 100 pounds.

He was very anxious to learn, so I turned him over to Jean-Paul, a burned-out French-Canadian ex-logger who was tougher than nails. Surprisingly, they made a good pair.

Otto was born in 1938 in an orphanage in Hamburg. His parents had abandoned him. The orphanage burned down during the war, and at six years old Otto was turned out onto the streets, eating out of garbage cans and surviving as best he could. He had absolutely—and I mean absolutely—no morals. Nobody had taught him any values whatsoever. All he had learned was basic writing, which was a miracle. Otto thought nothing of eating someone's lunch if he was hungry, or taking someone's personal tools or clothing. After some of the stealing stunts that he pulled off, a few of the other men became mad enough to beat him up.

However, Otto looked up to Jean-Paul, who was a middle-aged man and a father figure. Jean-Paul would talk to him, and if he felt Otto wasn't getting the point, he'd haul off and cuff him across the ear.

In the early days, Otto would look at you with those great big blue eyes, never hearing a thing you said. After three years, however, a new mature look came into his eyes, and you could actually sit down and talk to him for a while before his attention started to wander. He eventually became a very hard-working painter. He was totally fearless, with no concern for threats, police, authority or heights.

After three years he wrote to the city of San Francisco, applying for a job as a painter on the Golden Gate Bridge. This was a dream of his and, surprisingly, he was accepted. Jean-Paul acted as if he'd lost a son when Otto left for the United States. I hope Otto learned to respect others, and I hope he had a good life.

Drop everything

One day in late fall, when there was over a foot of snow on the ground, I finished work and hurried to put on my winter clothing for the long walk to the security gatehouse in the dark. As usual there were 400 to 500 workers strung out along the road to the gate, where some of us would be randomly checked for stolen items.

We were suddenly blinded when three giant searchlights atop the security building were switched on above us. Men stumbled into each other, trying to shield their eyes from the overwhelming light.

What in hell was going on?

I went through the gate without being searched and carried on to the parking lot and the bus I drove between the plant and Terrace (more on this in Chapter 7). I knew if I waited long enough, one of the men coming out would know what was going on.

Sure enough, one blabbermouth knew what all the lights were for and proceeded to tell everybody. Apparently a National Film Board film crew was doing a feature on the company's decision to go ahead with the secondary power tunnel to the Kemano powerhouse. They were getting some footage of all the men coming out of the plant. We accepted this story, and everybody snuggled down for the one-hour drive home to Terrace.

The next morning was a very different story. We arrived at the plant just as it was getting light. We stumbled through the security gate, and what to our wondering eyes did appear, spread out on the snow for more than 300 feet from the gatehouse, but bundles of paper towels, toilet paper, light bulbs, rolls of tape, work gloves, wrenches, pliers, hammers and nearly anything that wasn't nailed down.

The spotlights the previous night could have been a snap body search, and well over 150 items had been dropped in the snow.

Extrapolate that haul into possible routine theft every day, and you are talking real money.

Poor Owen

With all the thievery going on in the plant, the company decided to set up a Salvage Coordinator. If you wanted to buy anything in the plant that was lying around unused or surplus, you could negotiate for it with the coordinator. Employees were generally charged a nominal fee. They paid the coordinator, and he gave them a detailed pass to get the item out through the security gate.

This looked like a good job for an old-timer who had served his time in the smoke of the potlines. It would also be a steady dayshift job, which was much desired by all the shift workers.

(As an aside, when I look back over my thirty-six years at Alcan, I can see that men who worked in the potlines, doing shift work, until they were fifty could no longer sleep well because of the constant shift changes. They were not happy anymore. In some cases their health broke down and they ended up on long-term disability. Many developed Alzheimer's, while others got disgusted and quit, losing their pensions.)

The winner of the coveted Salvage Coordinator position was Owen, a tall lanky fellow, very good-natured, who had served his time in the potlines and got along with everybody. I think he got to know every one of the 1,800 people in the plant, by one means or another, over his many years with Alcan.

Everything went along fine for him for a couple of years. Then suddenly Owen was fired.

In all my years, I knew of very few actual firings of staff members. If you were on staff, you practically had to commit murder to be fired. So the rumour mills for both staff and hourly employees took off, grinding at full speed day in and day out. Owen didn't have an enemy, so everybody was pulling for him. The odd sorehead kept his mouth shut because Owen had that many good friends.

Slowly the mills ground, information dribbled in and the truth finally came out.
 a. Owen was too easy-going and good-natured to charge the proper price for anything.
 b. Other people were being looked into.
 c. Nearly everybody who knew him couldn't imagine he would do anything crooked. We figured the problem was probably the opposite: he was far too easy when it came to charging people.

Rumours were that he would lose his pension, which was pretty disastrous news. How could they take his pension away? He had put in enough time in the potlines to get at least a small pension.

His wife was a union member and worked in the plant. According to rumour, she went to the president of the union and he gave her the address of a lawyer the union had used. The lawyer allegedly went to the managers and explained that he had taken all of Owen's salvage passes and gone through them carefully. When he traced who had taken advantage of Owen's good nature, it turned out many of the offenders were members of upper staff. If the company had fired Owen over this issue, then they had to fire the staff members. We were not sure if this version of events was true or not, but Owen was reinstated, with full pension, and signed a secret agreement for the compensation he received.

Salvage pass

Another day I was in a lineup of five cars, waiting to leave the plant through the main security gate, but none of the vehicles were moving. I noticed people running around frantically beside the gatehouse, so I pulled over and parked, then walked up to the gate, thinking maybe I could help fix whatever the problem was.

Right beside the gatehouse was an interesting site. A car was pulling an old broken-down trailer loaded with plywood. The axle on the trailer had broken. The best solution, to avoid the lineup, would have been to get back in the car, drag the trailer through the gate, pull over to the side of the road and remove the load.

Instead, for some weird reason, they were taking the large pile of plywood off the trailer, one sheet at a time. I noticed that the first two sheets they took off were old, dirty, 3/8-inch unsanded plywood, but the other twenty-four sheets were new, 1/2-inch, sanded plywood.

William, the owner of the truck and trailer, was frantically waving his salvage receipt at the chief of security.

I gave them a hand, and in a few minutes the trailer was empty so they could pull it out of the way without damaging the pavement. I went on my merry way, out to lunch uptown.

The next day the whole sordid story was all over the plant.

Pot room line 7 had been mothballed for many years, since the 1971 recession. Stores had been using the large empty building for covered storage. Now line 7 was going to be put back into production, so they were taking down all the 3/8-inch plywood that they had used as temporary walls to protect the equipment being stored there.

All these many sheets of rough plywood became surplus. William had heard about this, and he paid two dollars each for twenty-four sheets of 3/8-inch rough, used plywood. Apparently he had been allowed to pull his trailer right into the building where all the old plywood was stored.

Right beside the pile of surplus plywood was a pile of nice new, clean, 1/2-inch, sanded plywood. Greed took over, and William piled sixty sheets of new plywood on his trailer, covering it with two sheets of 3/8-inch rough plywood. With bad luck, the trailer axle broke just as he was going through the security gate. When the security guys were helping him unload the trailer, they discovered the switch.

William lost his job as foreman, plus his house, and he left town in disgrace. It still doesn't make sense to me that people would risk everything for a few dollars.

Do you really know anybody?

I was a casual, "smelter type" friend with Louis. We knew each other from the plant, and for a period of time we worked together. I really thought I knew him.

He was much older than me, so he retired long before I did, in 1980, after working for the company for twenty-six years. When the time came, he asked me to help him clear out his office, taking most of his belongings through the security gate for him the day before his official retirement. This was because I had a car gate pass, so I could drive the load of items from the office building to his car in the parking lot. For some reason Louis didn't have a car gate pass.

The day before his retirement, we went to our office and loaded all his personal belongings in my car. When we got to the gate, I rolled down the window and told the security officer that we were taking Louis's personal stuff out to his car because he was retiring. The officer wished Louis a happy retirement and waved us through.

We were loading everything into Louis's car when his lunch bucket fell open in my hands. It was crammed full with two light bulbs and numerous packages of toilet paper.

I was dumbfounded and took a hard look at Louis. He stared back at me and said, with a great deal of malice, "Don't look so surprised. I hate this bloody company. Furthermore, in twenty-six years I have never bought a light bulb or any toilet paper."

He didn't seem to consider that he had just jeopardized my gate pass and my job. I was very disappointed in him. This was a fellow who I had taken out of the plant, in my car, for lunch uptown, probably more than a hundred times. I thought I knew him, but this built-up, hidden anger was totally new to me.

On reflection, though, I realized that whenever we came back from lunch uptown, he always had a reason to go to his car in the outside parking lot and then walk in from there. How many of those hundred times had he jeopardized my car and job?

Nevertheless, we gave him a good retirement party, and he was the perfect, affable guest of honour.

A few years later I was in the town he had moved to on holidays, and on the spur of the moment I decided to look up "good old Louis" and see how he was doing in retirement. I pulled up to his house with Mary in the car beside me. There was Louis, up on a ladder, washing windows on his house.

I rolled down my window and called to him, "Hi, Louis! It's Alan."

Then I got out of the car, glad to see him so healthy. I started to walk up the driveway, and Louis came down a few steps on the ladder and stopped.

He looked at me, with a very angry expression, and yelled, "What the hell are you doing here? I don't ever want to see you, or anyone else from Kitimat, ever again. Now just get in your car and leave."

In all the time we worked closely together, he had never ever showed anger or frustration. Anyway, I left, sadly, and my memory of him is now permanently altered and is directly related to ass wipe.

WCB malingerers

Theft doesn't always involve material goods.

In 1982 we got a new personnel manager at the smelter. As he settled into his job and started looking through files, he noticed a statistic that gave him concern. Over the previous few years we'd had an average of over eighty people off work on either short-term or long-term disability at any one time.

It really destroys the morale of working people when they see their so-called fellow workers wandering aimlessly around town every day. Especially when some of them are boasting about how they fooled the personnel department, the medical system and Workmen's Compensation Board (WCB) people.

We had two deadbeats I know. One got a small cut—and I mean small—on his wrist and ended up on WCB for over a year.

Another person had developed a bad back and boasted that he was never going to return to work ever again. He never did. He was a great actor.

Supercop was taken off regular security to check out all these people on disability. He asked the personnel department for their names and addresses, and also requested a powerful set of binoculars and a telephoto movie camera.

He would go out at any time of the day or night and watch the homes of suspects, taking movies of them doing amazing things, such as a man with a bad back getting on his motorcycle, accidentally dropping it, then bending over and picking up the 850-pound bike.

Another fellow with a bad back was caught building his house alone, manoeuvring the 400-pound centre roof beam into place by himself. Another bad back was caught doing the rumba at a local dance, thoroughly enjoying himself. The shenanigans went on and on. All in spectacular colour TV.

After Supercop gathered all the evidence, which took less than three months, each person was called in, shown the relevant video and told to be back at work the next day. If they didn't show up, they would be fired. Everybody knew what had happened, and some of these guys took an awful ribbing when they got back to work.

If I remember right, they got sixty of the eighty back to work. As well, word of what happened spread throughout the plant, so all of us were good little boys, for a couple of years at least.

Shot on the Rise

When I walked into my building trades maintenance planner's office one morning, I could find nowhere to sit. Planted in all the chairs were three foremen and two gang leaders, all talking at once. As I pieced the conversation together, I figured out that somebody had been shot the previous night. When I started picking up the details, I realized that with all the single man and boarders in town, it was a miracle that something like this hadn't happened long ago.

Apparently a husband had become suspicious that his boarder was playing around with his wife. The husband was on afternoon shift, but the previous day he had checked out after only four hours and gone home early. He sneaked up to his house and peered through the boarder's window. There was his wife, in bed with the boarder, who was lying on top of her. The husband had come prepared. He gently opened the window far enough to stick his rifle in, then shot the boarder through the chest, in one side and out the other, killing him dead.

About this time our general foreman came through the door. "What the hell is going on?" he asked. "Isn't anybody working?"

Sandy, one of the foremen, spoke up and said, "Barney from Casting caught his boarder in bed with his wife and shot him right in the act."

The general foreman, without missing a beat, said, "Well I hope he was a true sportsman about it and shot him on the rise."

Sandy, the designated spokesman, asked, "What do you mean, about being a sportsman and shooting him on the rise?"

The general foreman snappily replied, "Any good duck hunter who is a true sportsman doesn't shoot ducks on the water, only when they are rising off the water."

Everybody now understood and had a good laugh, such is men's rough sense of humour.

CHAPTER FIVE

From Footloose to Financial Epiphany

Footloose Again   *107*
Fishing Eagle   *111*
The Two Rons   *113*
Killer Whale   *117*
My Floating Fiat Car   *117*
Epiphany   *120*
Revenge   *122*
Our California Baby   *123*
Gold Bricks   *124*

Joe and Sharon, 1963.

All through my youth I would get restless to see the other side of the mountain and would either stick out my thumb at the side of the road or jump a freight train. Since I knew how to shovel coal in a steam train engine, I would sometimes approach the fireman and volunteer to feed the boiler for him. He usually said yes, and away we would go.

As I look back on those youthful days, I realize my footloose ways must have driven my mother around the bend. I was usually considerate enough to leave her a note, but still—she had to worry about where I was, who I was with and when I might be coming home.

Footloose Again

So here I was in the 1960s, married with two kids, a house, two cars and a rocksteady salaried job. In 1963, Mary wanted to go home to her mum and dad's farm in Robson for the whole summer, with the two children. I only had three weeks' vacation time, so I elected to stay home and clean up the yard and the garage.

After a week of cleaning the restlessness bug bit me. I hadn't felt this way for over ten years. What could I do to satisfy it? I decided to take a packsack and bedroll and hit the road for two weeks. I had never been to the prairies, so I thought I'd try to get there. I phoned Mary to let her know what I was doing, and the next morning I went out on the road in front of our house in Kitimat.

What would our neighbours think, seeing me out there with rough clothes on and a full packsack? Maybe somebody I knew would pick me up. What kind of rumours would that generate in a small company town?

By nightfall I was 300 miles from home, on the side of the road in the middle of nowhere. Things were a little different now than when I was younger. I had a thick wallet full of money, and if I wanted to, I could afford a hotel room. No cars came along, so I took off into the bush, rolled out my sleeping bag and had a good long night's sleep.

I woke up and had a cold breakfast, but when I got to Vanderhoof I dropped in for two cups of coffee. I carried on to Prince George, but there was no road through directly to Edmonton, many long miles away. I had to turn and go northeast, all the way up to Dawson Creek, then way down to Edmonton.

I was very lucky on my third day out when I was picked up by Ray, a travelling Bible salesman who was heading home to New Hampshire. We got along great. What really bonded us was my helping him pack his heavy battery for two miles through wet prairie mud to the nearest garage when the generator/battery quit just outside Whitecourt, Alberta. We got the battery charged and, eventually, the generator repaired, and spent the night in a motel, eating restaurant food. Ray could not figure me out, and we ended up talking about each other's lives and the Bible. I bought one of his Bibles for twenty dollars. It became our family Bible, with all the births and deaths recorded in it over the years, and I treasure it.

Ray ended up asking me to come with him to New Hampshire to meet his family. I figured out that I could make it to New Hampshire and back home within my two weeks, the Good Lord willing. If not, I would take the bus home.

East Coast USA, here we come!

We drove south through Edmonton and Calgary, talking all the while, and ended the day in a little jerkwater town just north of the border at Coutts, Alberta.

The next morning, after breakfast, we approached the US customs agents. They asked to see identification, and the inspectors found out that I was a Canadian. Ray, not wanting to create a problem for himself, blurted out that I was a hitchhiker who he didn't know. The inspector ordered me out of the car and back to Canada customs. Goodbye, Ray and New Hampshire.

Disappointed in Ray's behaviour, I hitchhiked back up to the Trans-Canada Highway and stuck out my thumb to go east. I got a ride in a great big loaded car carrier heading to Lethbridge, Alberta. Chatting away with the truck driver, Jerry, I found, much to my surprise, that he was only sixteen years old. He was a big beefy guy but still had a bit of baby fat around the face.

He had six new cars on board and had to drop off two in Lethbridge. He said this would take only a couple of hours. Then he'd be back on the highway to Winnipeg. He offered to pay me ten dollars to help him unload the two cars, so for the fun of it I agreed.

He drove that big rig like a maniac. How he had survived until then, I don't know, but approaching the outskirts of Lethbridge we rounded a corner at a terrific speed. Right in front of us was a railroad underpass with a low clearance sign. Jerry was going much too fast to stop, and the two top cars had their roofs pushed in. Even with all the usual racket of the car carrier, we could hear and feel the cars being hit. This didn't seem to bother Jerry one bit. Without stopping to assess the damage, he just sailed into the dealership, jumped out with his papers and went into the office.

Out came the manager, who was not a happy camper. Jerry hadn't a care in the world as the manager examined all the damage. We unloaded the cars and went merrily on our way. We crossed the Alberta/Saskatchewan border just before nightfall, and I asked to be let off in the middle of a hayfield. Jerry was planning to boot it down the road for a few more hours before pulling over and sleeping in his cab.

I had never been to the prairies, so it was a weird feeling to get out of the truck in the middle of nothing. The grain had been harvested, so all that was on the flat ground was four-inch stubble—no trees, no fences and no bushes to hide in, and not even a barn to sleep in. Just endless prairie as far as the eye could see.

I walked about a hundred yards into a field and unrolled my sleeping bag. The sun was slowly setting, and I felt like the whole world could see me getting ready for bed.

At this moment the reality of where I was and what I was doing hit me. What motivated me to leave my nice home and comfort and get out on the road again?

I fell asleep quickly and awoke just as fast.

Something strange had woken me up, but I didn't know what.

All I could see was total blackness. Looking closer, though, I saw seven or eight lights coming slowly toward me over the field. It looked like a bunch of people with flashlights trying to find me. They came closer and closer, until they were right in front of my face.

Fireflies!

This was something I had never seen before. What a relief.

I settled down and went back to sleep, still wondering what had awoken me when the fireflies were so far away.

In the morning I woke up to the great big sun glaring straight in my face. I got up, but with no firewood I couldn't even cook breakfast. With nothing else to do, I packed up, then looked at my watch.

Christ! It was only five o'clock! What kind of country had I come to where the sun was blazing at 5 a.m.? Anyway, I was up, so I decided I might as well get going. I staggered out to the road and in no time flat had a ride in a cattle truck, loaded with steers for the slaughterhouse.

The driver was heading all the way to Winnipeg, and we journeyed together to that city. During the long afternoon, as we were heading east, we could see a great black cloud straight ahead of us. As we got closer, lightning bolts were visible, zigzagging continually from cloud to earth.

Bob, the driver, casually said, "Well, I guess we're in for it."

I asked him what he meant, and he said, "We have thirty head of cattle loose in the truck. When lightning strikes close by on one side of the truck, they will all suddenly move to the other side of the truck. Under the right conditions, it can turn the truck over."

Now that was a nice thought as we roared down the road at 60 miles per hour, straight at a monstrous black rain and lightning storm. At three in the afternoon it got darker and darker until you could only see about a hundred yards ahead, and the rain started coming down in buckets.

At first the lightning was hitting the ground way out in the fields. Next thing I knew it hit a fence post on my side of the road, shooting along the barbed wire on the fence posts. The thunderclap sounded like ten sticks of dynamite going off.

All the cattle lurched to the other side of the box. Bob had his work cut out for him. First there was a violent turn to the left to keep the truck on all twenty-six tires. After it settled we were on the left-hand side of the highway, nearly in the ditch. Thank God there wasn't another vehicle coming toward us. Bob slowly returned the truck to the right side of the highway. He didn't seem fazed by the lurching, although he did slow down a bit. Throughout the next half hour, as we passed under the thundercloud, the same situation happened about ten times. It sure terrified me!

We got to the big city of Winnipeg, and Bob was kind enough to drop me off at a hotel before he headed for the cattle yard in the pouring rain. The next morning was hot and muggy. The humidity was unbelievable, with 90° heat. I wandered around the city for the day.

Then I made up my mind: I'd had enough adventure and I wanted to head back home. Maybe, just maybe, married life had softened me up, or age had made me wiser.

I decided to gain something from the trip, so I bought a nice 1948 Mercury coupe in good condition, with no BC rust, for $425. I drove it back home, cleaned it up, threw a coat of paint on it and sold it for $900. This made me feel better because the trip expenses were paid for, and I decided to retire my rambling thumb.

On further reflection, I decided that routine and responsibility must've gotten to me, and I'd just had to break loose for a while.

Fishing Eagle

Luckily there were lots of places near Kitimat where I could go to get away from routine and responsibility for a few days. When the weather was just right and I was out in a boat on Douglas Channel, the world couldn't be better. You could look for miles in all directions and not see another person anywhere. No logging shows, no roads, just unspoiled wilderness. If you wanted to stop on a sandy beach, take all your clothes off and swim or sunbathe, help yourself. Maybe because I was born in British Columbia, I look at things a little differently than other folks.

I was never happier than when I was out on the water or in the wilderness, peacefully fishing or doing whatever I wanted.

On one such calm, sunny day, my son, Joe, and I were out in Douglas Channel, down at Jesse Falls, fishing for salmon in our 12-foot car-top boat.

Sometimes spring salmon will come up and school on the surface for a few minutes, then slowly sink. This attracted a lot of seagulls and bald eagles, hoping to get a feed.

We had picked up one fish when suddenly there was a splash about 30 feet away from the boat.

A huge bald eagle had his claws in a very large fish but was having trouble getting airborne again. We watched as he struggled valiantly, flapping his great wings in the water. You could see that he was losing the battle, getting wetter and wetter as he sank deeper and deeper. I'd heard stories of eagles and hawks sinking their claws so deeply in the fish that they couldn't release them and drowned. Now I was seeing it in real life.

Half the eagle's body was under the water, and ever so slowly he sank farther and farther until he could no longer flap his wings above the water. Now that fierce-looking head was the only thing showing above the surface. Then he was gone.

Joe and I sat there, trying to hold the boat in the same position, hoping the eagle would come back up close to us. The minutes seemed like hours, but finally something very black and wet broke the surface. He had finally gotten loose. He must have hooked onto a forty- or fifty-pound spring salmon, not one of those little three- to four-pound sockeye.

That eagle was sure a sorry sight, and he ever so slowly managed to lift his waterlogged wings above the surface and swim to the shore, about 60 yards away. He crawled up on land and shook himself all over, again and again. Then he held his wings out in the sun to dry off.

Joe and I returned to our fishing, but every now and again we would check on him. After about an hour in the sunshine he fluffed his plumage all over and away he went, none the worse for wear.

The Two Rons

Mary had two old friends with frustrated husbands. Both the husbands were named Ron. They had moved to the great northwest from the city with their rifles and hunting gear, and they wanted to get a moose. They had been hunting for over two years but had never even seen a moose. To me this seemed unbelievable. There were moose everywhere, especially in the area 200 miles east of us.

Mary made the mistake of boasting to these women, "My husband gets at least one moose every year. In fact, he's already got his moose for this year."

Well, one thing led to another, and in 1963 I ended up being dragooned to take these two city boys out to the wilderness, teach them the basics of moose hunting and get them at least one moose. I was not overly happy about this situation.

We decided to go hunting on the next long weekend, by which time, hopefully, the temperature would have dropped below zero and the bulls would be starting to rut. On a typical fall day, two or three bull moose will be peacefully grazing together, but when Jack Frost comes along that night, he somehow gives a huge boost to the bull moose's testosterone levels. The next morning they are fighting with each other because there are no cow moose present. They basically go sex crazy.

I told the two Rons that I would take care of the food and transportation providing they took care of their own personal needs, including sleeping bags. I also told them that I had an old friend, whose backwoods cabin I had been using for years. I could use his cabin until the snow flew. Then he would be using it for trapping.

My friend and I shared a common backwoods understanding of respect: "Use my cabin, but leave it in better shape than when you found it." After I used his cabin, I would always leave some tinned food so the pack rats couldn't get it. I'd also leave a bottle of rye whiskey in a special hiding place, fill the kindling box, replace the firewood I'd used and top up the coal oil lamps. All this common woodsman's practice I carefully explained to the two Rons.

We drove 200 miles into the interior and arrived at the cabin on a cold, clear night with a full moon and shining stars. The perfect beginning of a moose hunt. The cabin was quite a way from the gravel logging road, tucked into a little valley that ran up the side of the mountain looming behind the cabin. We got the stove going, I heated up the chili, and we were just sitting around when I got a bright idea.

I told the Rons, "Seeing that it's going to freeze tonight for the first time, the bull moose will be going into rut. So I'm going to go outside and try to call in a couple of bulls for tomorrow morning when we get up. That way we won't have to drag them through the brush after we shoot them."

They both looked at me like I had gone mad. Ron number one asked, "What is a rut?" So I explained to both of them all about the moose's sex life.

I'd no sooner finished than Ron number two asked, "How do you call a bull moose in?" Now I had to explain the process of calling a bull moose in for a challenge with another bull moose, as well as how to imitate a cow moose calling a bull to satisfy her sexual needs.

I went over to the grocery box, cut off a large chunk of cardboard and rolled it into a bullhorn to call a moose. Leaving both Rons completely confused, I went outside to issue a call of challenge to a bull moose.

What an evening! Full moon, sky loaded with stars and no wind, so it was deathly quiet. I listened for about ten minutes to see if there was any calling going on. Eventually I heard a bull very far off.

I was halfway through my first call when the door flew open and two heads popped out. They started laughing at me and at the unusual sound that I was making, loud enough to scare off any moose. I told them to be quiet and shut the bloody door, as I was waiting for a reply.

After waiting twenty minutes in the cold I made another call, and soon I heard a reply to the challenge. So I knew that the next morning we might have a visitor from up the mountain behind us.

I had just put my hand on the door latch when I heard another challenge call from a different direction. Great! We now had two bulls wanting to fight in our area.

I went inside, and the two clowns started making fun of my efforts, as if I was putting on a show for them.

I told them, "Tomorrow morning you better not go outside until I tell you. We'll make breakfast, put on all our hunting clothes, grab our guns and load them. Then we'll quietly—and I mean quietly—go outside, and hopefully we'll shoot one of the bull moose that I have called in. If you don't do exactly as I say, we will be going home. Hunting is serious business."

With that we had a small drink and went to bed.

Everybody was up bright and early. I put wood in the stove, and in fifteen minutes we were all warm and cozy, and all ablutions were done. I was elected cook, and as I was getting ready to prepare bacon, eggs and coffee, Ron number one said he wanted to go out to the biffy.

I said, "Okay, but take your rifle with you. There could be bull moose out there."

I was walking across the cabin floor to the stove with a carton of eggs when Ron number one opened the door and stepped back. The next thing I knew there were two explosions, then a third. By this time the eggs were on the floor, broken.

I looked out the door and there were two bull moose lying in the grass 50 feet from the door. Ron number one had made three beautiful shots. Breakfast was delayed while I bled them out.

Over breakfast, the two Rons had completely changed their tune. Now they had absolute faith in my ability to call bulls, even though the events of that morning were probably a one in a hundred chance.

The hunting trip was now over, and the work was to start. We had to drag the two 1,000-pound animals over to a large tree, hook up the block and tackle, lift the two moose up in the air for disemboweling, and then cut off the heads and feet and skin them.

Since we got our moose on the first day of a four-day weekend, we could let them hang for three days, which would help break down the meat and make it more tender. Bulls are usually much tougher than cows. I made the two Rons do the dirty work, and it gave me great satisfaction to see them both christened with blood, shit and hair. They were now nearly hunters.

The odd car came by along the abandoned logging road out front, and the hunters inside would stop when they saw the moose hanging from the branches. I let the two Rons take all of the attention as Great White Hunters.

After one car left, and they were all smiles and puffed-up chests, I deflated them by handing them each a shovel. Ron number two asked, "What's this for?"

I replied, "Dig a big hole, drag all the guts, legs, heads and hides in and bury them." With only one car, three men plus all our gear and more than a thousand pounds of moose was a big load, so I got them to do a little butchering to lighten the load.

We had only been home for a few days when friends started coming up to me on the street and at work, telling me all the gruesome details of our trip. I had repeatedly told the two Rons that this was my cherished secret hunting place, but now the whole town knew where it was. I could have cheerfully throttled both of them.

The next year I went back to the same cabin with one of my old hunting partners. You guessed it—the cabin was now a complete shambles. We put up a tent, but the traffic of hunters was beyond belief. We could have been in downtown Kitimat for all the people we knew who drove by.

Still, I had hunted here, and been successful, for over eight years, so I was going to give it a last try.

The next morning we went to my favourite spot, hoping there would be some good sign. We split up, and as I was going around a large spruce tree that had branches right down to the ground, I heard a crunching of leaves. I clicked the safety off my 30-06 Remington pump and slowly crept around the tree. Suddenly I was face to face with another hunter, each of us with a gun pointing at the other.

I am not a purist, but that finished my favourite hunting spot for me, thanks to the two Rons who couldn't keep their mouths shut.

Killer Whale

One nice sunny Saturday in June 1964, when the first of the spring salmon were coming up into Douglas Channel, Joe and I decided to go fishing in our newly purchased 18-foot boat with its 50 horsepower Mercury motor. Boy, were we ever proud of that boat.

We went all the way down to Jesse Falls and trolled for a couple of hours in front of the falls, but no luck. This was the only place I knew where there were delicious four- and five-pound rock cod, so I set us up for bottom fishing with rods, weights and bait. We had to stay in one place beside the underwater cliffs, so I also set out an anchor.

We broke out the lunch, and afterward, with the hot sun beating down, and the water just like glass, we got drowsy. I had my head over one side of the boat and my feet over the other, and with the gentle rocking of the boat, I fell fast asleep.

I awoke suddenly when two unusual things happened at the same time. There was a strange, small rocking of the boat. And an unusual gentle but deep sighing sound, as gentle as an angel's kiss. I slowly opened my eyes and realized I was eyeball to eyeball with a bull killer whale. He was no more than three feet from me. I was looking at his huge eye, and his dorsal fin was towering more than six feet over Joe, who was still sleeping in the back of the boat.

He looked at me, and I looked at him for the longest time, with my heart in my mouth. One slow flick of his giant tail and we would be in the water, or eaten, take your choice.

Then he slowly sank down and slipped away, and I could nearly relax.

Believe me, when you are in a flimsy boat, a bull killer whale looks very big. Especially when you are only three feet away, eyeball to eyeball.

My Floating Fiat Car

I have always been a restless soul with too much energy, mental and physical. Over time, this has got me into some weird predicaments. The early winters in Kitimat were particularly brutal, with anywhere from 20 to 40 feet of snow keeping us inside a lot, and with me having nothing to do in my spare time.

I finally developed a philosophy that in order to keep myself happy and sane each winter, I would take on a large project. In the sixteen years that we lived in Kitimat I rebuilt ten wrecked cars and modified a 7 mm Mauser rifle with telescopic sights and a hand-built custom stock.

The little 1964 model Fiat 600 car I picked up for $1,000 in 1966 was nearly new but had been wrecked. I figured I could fix it up and use it to go back and forth to work, plus drive it for fishing and hunting trips. It would be cheap on gas and a good second car, and when I got it going in the spring of 1968, it ran better than I thought it would.

The spring salmon were running early that June, so I decided Joe and I would go fishing in our favourite hole on the Kitimat River. It had been raining heavily for a solid week, but had now slackened off, so away we went in the little Fiat. I expected the narrow abandoned logging road to the river would be muddy, but I had discovered, to my pleasure, that the rear-wheel-drive Fiat, with its motor over the rear wheels, would go through nearly anything.

We turned off the main road onto the one-mile stretch of ten-foot-wide logging trail. It was muddy, but not so bad. We got in about a quarter of a mile and were on a straight stretch, about 200 yards long, that appeared to be solid water. I stopped to look at the situation.

Joe was now thirteen years old and growing like a weed, nearly six feet tall, so I asked him to go ahead with his waders on, take a long stick and test the depth of the water as I slowly followed him in the car.

We had only gone about 50 yards in when I felt a strange sensation. If I touched the gas pedal, a sort of front-to-back rocking motion occurred. I was only going about one mile an hour, so I stopped and let Joe get way ahead of me. He was wading in water nearly up to his knees. That meant it was too deep and the motor was going to get swamped, right in the middle of nowhere.

I looked down at the floorboards, expecting to see some water, but they were dry. I couldn't figure out what was going on. I again put the car in gear and deliberately pushed the gas pedal down, then up. The front end came up, then down, and then the back came partially up. Lord love a duck! We were floating, and the doors were sealing the water from coming in.

I pulled my waders on, got out of the car and opened the rear engine hood. I wanted to figure out why the fan wasn't sucking or blowing water around to kill the engine. Turns out there was a sheet-metal plate below the engine compartment, sealing it all up tight. There was also a manually operated vent flap valve—which was, luckily, closed—that sent the radiator heat either into the passenger compartment or out the bottom.

I jumped back in the car and signalled Joe to get in too. Away we went, boating up the road. The winter tires pushed us along at about five miles per hour.

I couldn't wait until Monday morning. I was going to pull a joke on my fellow planner, Alec. Every lunch hour we would take my car and drive down to the ocean, where we would sit in the car on the seaplane ramp, watching all the wildlife and eating our lunch. Just 200 yards from the ramp was deep water, with three large dolphins, or mooring posts, in a row where seagoing ships could tie up.

On this Monday, when we arrived at lunch hour the tide was in, and three-quarters of the concrete ramp was underwater. We sat there quietly eating our lunch. When we were both finished, I checked that the doors were closed and the windows open, just in case we had to abandon ship. I also made sure the heat vent was closed to prevent water pouring in.

I looked out toward the dolphins and said, "I think I saw a sea lion out by the dolphins. Let's go and check him out."

Alec stared at me with a horrified look on his face and said, "Are you crazy? You can't go out there! This is a car, not a boat."

So I said, "Well, let's go just a little closer. Get a better look." I started the motor and slowly drove the car forward till it was in knee-deep water and started rocking.

Now he couldn't get out, so I said to Alec, "You know, I think this car will float. Let's try it out. You might want to open your window, just in case."

By this time he was looking out the side window, with the water six inches up the outside of the door, screaming, "You're crazy! This car will sink like a stone. Let me out. Back up so I can get out."

I gave the gas a nudge, and we were floating. I said, "See, we're floating and there is no water coming in."

He yelled, "Maybe so, but the water out there is 40 feet deep."

I replied, "Aw, come on, Alec. Let's give this little floating car a try."

He didn't say anything, so we cruised out and went around one of the dolphins, then back up the ramp.

Alec yelled, "You rotten son of a bitch! You knew this car wouldn't sink. You nearly gave me a heart attack."

Epiphany

If you looked really quickly at us as a family, it appeared we were doing pretty well: we had two cars, one of them a nice big Cadillac; a 14-foot Shasta trailer; some bonds in the bank; and a month's holiday every year. I had an inkling at the back of my mind that something was not quite right, but on the surface, when compared to others, we were doing great.

In 1964 we were on holiday, visiting Mary's family down in Robson, BC. Our plan was to loop down into the United States with the trailer and work our way home to Kitimat. Ever since coming north I had developed a hang-up. Every year I just had to get out of the north, even though it was a thousand-mile loop on unpaved roads to Vancouver or the Okanagan. The roads were getting better each year, but it was still a rough trip. And I still had to get out for at least a few weeks.

When we looped down into Washington state this year, the Caddy broke down 20 miles north of Colville, Washington. We had to have the car and trailer towed into town—the trailer went to a trailer park, and the Caddy to the GM dealer.

We had blown the motor, and a rebuilt unit would cost $600. I didn't have that kind of money in my savings account, although I did have $5,000 in bonds. I phoned my banker and explained my predicament as if it were no big deal, but he gave me the gears. Even though the bonds were locked up in our safety deposit box, he wouldn't send me the money as a short-term loan. We talked, we argued and finally I had to beg him for a lousy $600 loan at 10 percent interest until I got home. After much preaching on his part, he finally agreed to send me the money, which I would replace when we got home.

I walked out of that phone booth a mad, changed man. Never again was any son-of-a-bitch banker going to make me beg for money. I didn't know how I was going to accomplish this, but I sure as hell was going to find out.

The car was finally fixed, and we headed north for the 1,400-mile trip home. This drive gave me a lot of time to think. I decided that, seeing as how I hadn't a clue about handling money, I was going to start by reading some books on money management.

As soon as we got home I went to our local library and checked out two books on managing household money. After I read them I went off the deep end, checking out a book on how to start and run a profitable small-loan business. That wasn't really my cup of tea, but I learned a lot of things. I then picked up a book on the laws of marketing. This steered me into buying a complete course of four books on real estate investment, and another book on commercial property appraisal. All this reading took nearly a year, but in 1966 I was ready to dip my toe into the pool and acquire some commercial real estate.

I talked to my hard-nosed banker, and he suggested I set up a commercial account with him, then go to an accountant and set up a limited company. I went to Carlyle Shepherd Accountants, and were they ever pleased to set up a company long before the first dollar was spent. They had never done this before, and they did a beautiful job that lasted me over thirty-one years.

Now the time came to put in practice what I had learned. I had $6,000 in bonds, but when I checked out what potential rental properties were available in Kitimat, which was practically a company town,

I discovered you had to have a lot more than $6,000 to get started. So I drove the 40 miles to Terrace and latched on to an older salesman who knew Terrace inside and out. I told him that I was looking for some form of multiple rental property. My small amount for a down payment restricted what was available, so I told him I was in no hurry. If something came up that I could repair, and that the owner was desperate to sell for some reason, he should let me know.

Three months later he contacted me about four duplexes that were owned by three different people. They were fighting among themselves and wanted to sell the properties and split the proceeds. I came out with Mary and looked them over. All I could see was work and more work. Mary and I talked it over and determined that it would take at least a year, working every weekend, to get everything in shape. But then again, I was never scared of work.

Mary convinced me that purchasing the properties would be taking the first step to financial independence. So we offered a down payment of $5,000 and signed the Agreement of Sale. Six years later we had twenty-four rental units, and we were starting to see a bit of money left over for us after monthly payments. I could easily fill two books on the adventures we had as novice landlords, not to mention the steep learning curve for me as I had to ensure I kept my property management completely separate from the rest of my life as a member of management with Alcan.

Twenty-five long years later we had everything paid off. Over the next five years we slowly sold it all and retired. And I never had to grovel to a banker again.

Revenge

By 1970 we had nearly paid off the mortgage on the original eight rental units in Terrace. According to the detailed course in real estate investment I had taken, I should never have a piece of property paid off because that meant I was reducing my equity profits. With this in mind, I wanted to refinance and buy more revenue property. Mary had spotted a fourteen-unit motel in Terrace that was up for a distress sale at a ridiculously low price. It was rundown but had good location and potential.

I went to my dreaded banker in Kitimat with a detailed description of what I wanted to do. He looked at my plan and proceeded to demand security after security until he had tied up everything I owned. This was four times the value of the motel, and he wanted to finance the purchase at a ridiculously high interest rate.

However, he was no longer dealing with the gullible schmuck of six years earlier. I took his paperwork over to the Bank of Montreal and explained the situation to a banker there. Then I said I would move all my accounts over to his bank if he would give me a mortgage on the motel and leave the rest of my holdings free.

The BoM banker laughed and said, "No problem." We worked out all the details, with a very good interest rate.

Then he got a twinkle in his eyes and said, "Do you mind staying here while I phone Mr. X and give him the gears? You can listen in on my other phone if you like."

He picked up the phone and dialed my former banker, and I heard a side of banking that I didn't know existed. I assumed from this that bankers must get together to discuss business, and out of that comes a certain amount of good-natured rivalry.

My new banker was really ruthless with my old banker, laughing and kidding him over what he had tried to do with me. He ended up by telling him that he would have somebody on his doorstep the next morning with authorization to pick up all my bonds and accounts.

Our California Baby

We were on a long, extended holiday in California when, at the age of thirty-six, Mary announced that she was pregnant with our third child. She was not a happy camper, but we were thrilled when our "California baby," Skye Morgan McGowan was born.

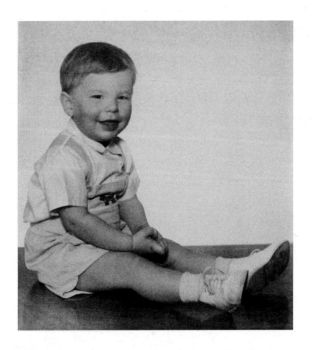

Skye McGowan.

Joe was now ten, and Sharon was nine, and between us, we spoiled Skye all to bits.

Gold Bricks

I was finally interested in making some money, so when I read about a teacher at the local high school who was going to give a two-month night school class on international finance, I decided to sign up. This was a little out of my league, but I figured now that I owned some rental property, I might get something helpful out of it.

As I expected, most of it was over my head. But when the teacher started explaining why gold was thirty-four dollars per ounce, I felt like I was on familiar ground. I had spent a lot of time in mining during my early working life, and I had learned that there was a thriving underground economy and big business in bootleg gold. This untraceable gold was bought and sold at considerably more than the international figure.

The teacher stated that within one year the restriction on buying and selling gold was going to be lifted, and gold would rise to its proper value. He predicted that it would float up to nearly a thousand dollars within a few years, so he advised us to buy all we could at the present deflated price. He was so sure of this prediction that he personally was going to take a second mortgage on his house and buy more gold.

I went home all enthused, but then reality kicked in. Mary, with her accounting skills and as treasurer of our rental company, "Skeena Estates Limited," explained to me that we were already deeply in debt, what with the rental properties we had acquired in Terrace and various mortgage payments. As well, because of a small recession in the logging industry around the northwest, we were down to a 50-percent occupancy rate. Each month we were taking a small amount out of our personal savings to pay the mortgage. Not a very good financial position.

This was frustrating for me, but having been in the rental business for four years, I had learned one ironclad lesson: When it comes to money, do what your accountant says.

At work nearly all of my major projects had been cancelled, so I had been transferred to Pot Room Maintenance. This was a bit of a comedown for me, to be stuck at a desk, firing out paperwork, but I was grateful to have a job. More than sixty people had been let go.

Each day at coffee break we would all get together to discuss the work, and then there would be a free-for-all, when everything would be discussed. This included Greg, the general foreman, and Ole and Russ, the maintenance foremen. I was only just getting to know these three gentlemen, so most days I was keeping my mouth shut as much as possible so I could learn all the ins and outs of doing maintenance for all the pots in the smelter. We had forty-eight men doing this work, spread throughout the whole plant.

After coffee in the morning, to my surprise, Greg, Ole and Russ would have a ten-minute discussion about buying and selling on the stock market. I was very impressed with their knowledge and their active participation. Over a period of time I learned a bit about the stock market.

Getting up my courage one day, I thought I would contribute my two bits worth and tell them what I had learned about gold. I had just started to describe the class that I had attended and the teacher's comments about gold bullion, and I had everybody's attention, when John, one of our maintenance man, came halfway through the door. He stopped when he saw all of us.

Big John, was a powerfully built, middle-aged immigrant from Poland. An ex-professional wrestler, he hardly ever spoke and, for some inexplicable reason, was totally intimidated by management people.

Ole saw him standing there, leaning against the door frame, and quietly said to him, "Just a minute, John, and we will get to you. Okay, Alan, continue."

I briefly explained what the night school instructor had said, and repeated his prediction about gold going way up in price.

Ole then said, "Okay, John, what can I do for you?" As they discussed what work Ole wanted him to do in the pot rooms, the rest of us went back to work.

Let's turn the clock forward a few years. I looked up the price of gold, which was pegged at $750 an ounce. I hadn't bought an ounce, but I was still anxious to talk about it and how close I had come to making a pile of money on gold certificates.

The next day at coffee break, the same people were gathered except for John. The business discussion was finished, and so was coffee, and I was ready to raise the topic of gold.

Then, déjà vu, enter Big John, standing in the doorway.

I asked my burning question: "Well, it's been a few years since we had that discussion about buying gold bullion or certificates. Did any of you do anything? Ole, what about you?"

He matter-of-factly replied, "Based on what you said, I bought $2,000 worth of gold certificates, and I'm very pleased with your advice."

I turned to Russ and asked, "Russ, did you also buy some gold?"

He replied, "I have no use for gold, but I bought $10,000 worth of silver bullion, as all of you well know."

Three weeks after our first talk about gold, we were all sitting in our respective offices, three in a row, with glass partitions between them, when Russ, in the centre office, received a phone call. We each could hear his wife screaming at him, even over the drone of the huge pot room fans.

Russ was pleading with her to stay calm. Then he said, "I will come home and straighten it all out." Without another word he slammed down the phone and rushed out of his office. We didn't see or hear from him for two days.

He returned to work the same old happy-go-lucky Russ. We had a hard time waiting for the 10 o'clock coffee break to hear what all the fuss was about. Finally, cups of coffee in hand, we went to Ole's office, biding our time until Russ made his appearance. When he walked through the door, Greg nailed him. "Russ, you took out of here two days ago without a word. That's not like you. We're all a team here, and with a little communication we would have covered for you with your crew. Now what the heck is going on?"

Russ blurted out, "It's all Alan's fault! He told us about the embargo being lifted on trading in gold and that we could make a fortune. I read about those two brothers in Texas who said they were going to corner the silver market and make a fortune. So I rushed out and bought $10,000 worth of silver. I don't like gold. I thought I was buying silver certificates, not silver bullion. Boy, was I mistaken."

After a large gulp of coffee, he continued. "Two days ago a large truck pulled up to our house, loaded with tons of little silver bars. My wife had a fit! What to do with all these thousands of silver bars? That was the reason I had to rush home. We ended up piling them in our bedroom downstairs."

Ole jumped in and said, "What in the world are you going to do with all those silver ingots?"

Russ smugly looked around and said, "Oh, I've already taken care of that problem. That's what I was doing all of yesterday."

Greg asked, "What brilliant solution did you come up with?"

Russ replied, "Well, it really wasn't that much of a problem. I laid them all out flat on the floor of the master bedroom, then covered them with interlocking sheets of three-quarter-inch plywood. Then I covered the whole works with wall-to-wall carpeting. The one small problem that I had was I had to cut three inches off the bottom of the door."

We all had a good laugh at Russ's ingenuity, but that was typical of him.

Many years later, when the price of silver, finally went way up, Russ sold the silver for a reasonable profit, and a Brinks truck came to town to take all the silver bars to the new owner.

But getting back to the question of gold, I turned to Big John, standing in the doorway, and said, "If I remember correctly, John, when we had that original discussion ago about gold going up in price, you were waiting to see Ole about something to do with work and heard every word that we said. Did you by chance do anything about it?"

John looked at me and quietly said, "All you bosses, very smart. I listen very carefully. You invest, so Big John invest."

I replied, "John, I know it is none of my business, and you don't have to tell us, and we would understand, but did you buy some gold?"

John replied, "Yes, I bought $30,000 worth of gold certificates."

You could have heard a pin drop. Then everybody exclaimed, "Good for you, John!" He got his work orders and was gone.

Ole pulled out his trusty calculator and determined that if John sold that day, with the price at $750 per ounce, his profit would be $631,500. We were all dumbfounded by Big John's casual approach to a fortune.

And me, with all my fancy talk, I didn't buy one ounce of gold and looked like a big buffoon.

CHAPTER SIX

Major Project Coordination Planner

The Wet Scrubbers *131*
Ask an Expert *135*
A Mighty Big Lift *140*
The Double Switcheroo *168*

One of the eight giant wet scrubbers.

In 1971 I was promoted to a newly established position with the very fancy title Major Project Coordination Planner.

Not to bore you with too many details, my new job was to take on an upcoming large project, break the project down into its basic elements, and then, with the input of all the various trades concerned, develop a detailed critical path schedule and coordinate the ongoing project. For example, one project might be shutting down the whole large casting centre, with two casting furnaces and the casting pit, at a shutdown cost of $2,000 per hour, so we could rebuild the centre in the shortest possible time.

Being the new boy, in a new capacity, I had to handle everything with a great deal of tact, as there were many staff people, including a foreman and general foreman with easily bruised egos, who were not totally convinced that this new concept would work or was necessary. Since I had to extract information and cooperation from numerous people to coordinate major projects, I knew I would have to rely on my philosophy of "sliding further on honey than on sand" more than ever.

The Wet Scrubbers

With a McGill University course on major project coordination planning under my belt, I was given a nice office upstairs in the maintenance building. I also took some night school courses, with the result that I was allowed to join the Society of Architectural Engineering as an engineering technician, with a nice big membership document to hang on my office wall beside my certificate from the McGill course.

My boss immediately gave me my first project. I was to develop a detailed plan to overhaul the eight wet scrubbers throughout the potlines. These were monster red cedar towers—15 feet in diameter and 60 feet high—each spraying 500 gallons of water per hour on hot, dirty, exhaust gases from the potlines that were pushed up the towers by the wind of 100-horsepower fans. The water spray cleaned the gases out of the pot room, but it generated a sticky black goo that was 5 percent hydrofluoric acid. This was dumped in a settling pond, and eventually some of it found its way into the yacht basin and the ocean. (An interesting side note: In the first year of the plant's operation, an ocean-going vessel from Prince Rupert happened to anchor in the bay for two or three days. When it was ready to leave, eureka! A magic fairy had removed all the barnacles. It didn't take long for sailors up and down the coast to learn that you could get your hull cleaned in Kitimat for nothing. This went on for years, with or without the company's knowledge, until 1975, when we started to put in dry scrubbers and shut down the wet scrubbers. You could call it a goodwill gesture by Alcan, cleaning all those hulls for twenty years without charging for it.)

This project was right up my alley, as I had experience with all phases of the work. And it was definitely time for the overhaul. After sixteen years of operation in the highly corrosive atmosphere, the structural steel, the metal ductwork and even the wooden towers had seen better days.

There was one drawback, a big one: If a wet scrubber is shut down for any amount of time, thick black fumes collect inside the pot rooms, where men have to work. This means no work on the pots can be done, resulting in a loss of metal production. Therefore, the pressure was on me to develop a plan to keep each shutdown for replacement and repair to a minimum, or the bosses would scream bloody murder.

Around this time I took two life-changing courses, sponsored by Alcan. They say that learning is a broadening experience that can transform your way of thinking and expose you to ideas you can benefit from. I agree with all this and have experienced it myself, but I also agree with people who say learning has a negative side as well.

First I took a three-week time and motion course, put on by the BC Research Council. We were taught how to solve workplace problems that at first glance seemed unsolvable. We learned to use different tried-and-true methods to analyze these problems and, amazingly, we would come up with a solution by thinking outside the box.

The next course taught me how to freethink in groups to solve difficult work problems. This lasted two weeks and made me respect, and appreciate, those oddball, freethinking people within the company. These people are usually treated with disdain because they don't fit in. The result is that they become isolated and are fired or quit, even though they may have a lot to offer their employer. One thing I learned after working for a large company of 60,000 employees for thirty-six years is that some of the biggest changes or inventions came from these freethinkers. In my various jobs with the company I looked for these people to give me radical ways of solving problems.

Both these courses were very exciting, almost intoxicating, and like a breath of fresh air for me. I had never been a team player, but they brought me into that orbit. I also became a lifelong adherent of the concept of "thinking outside the box."

The negative side of all this learning is that you become supercritical of the everyday, inefficient methods other people are using to do things that really are none of your business. There isn't enough time in your life to solve everybody else's work-related problems, so you have to learn how to leave them alone.

Getting back to my assignment to overhaul the wet scrubbers, I had to figure out "who had to do it," "how they had to do it" and "when they had to do it." To add to the difficulty, I had no authority to order anybody to do anything, just suggest. I dealt with all the trades with a velvet glove. I could go to the upper ranks, and they would remove the velvet glove. This was seldom necessary but was there if needed.

I developed a "critical plan," a layout of all the work actions that had to be taken to repair all eight wet scrubbers. I wrote all the steps out on a three-foot by twelve-foot piece of graph paper that I hung on my office wall. There were over seventy-five actions, spread over two months. This was not good: winter was coming, and all the actions were outdoors. Frankly, I didn't know how to overcome this problem.

With perfect timing, one of my many bosses came in to check on my first effort. Neil said, "Well, how is the planning coming along?"

I replied, "My theoretical planning is okay, but it takes too much time and we will end up waiting around in ice and snow."

Neil took a long look at my schedule, spelling out all the work for carpenters, millwrights, welders, plumbers and painters. Then he looked at me and asked, "Are all the different departments happy with the schedule?"

I replied, "Yes, they must be. They all agreed to the plan."

Neil said, "So they all agreed to the plan, but you are not happy because it's taking too much time?"

I considered this statement and mumbled, "Yeah, I guess so."

Neil replied, "To sum it all up, you have taken all these trades that work steady dayshift, and don't work Saturdays or Sundays, and made them all happy to do the job as they are used to, but you're not happy. Does that sum it up?"

Reluctantly I agreed.

Neil then said, "Alan, you're not here to try to make everybody happy at your expense. You're here to get a job done the best way you can. You lay out the entire project again to make you happy, and we will make sure it happens, come hell or high water."

With that he left, with me speechless.

I sat down and stared at the wall for a long time, trying to get my head around this new way of looking at things. I went home and thought about it some more, and finally I grasped the concept. I couldn't wait to get back to work and lay out the whole project my way.

End result: if we had teams working twenty-four hours per day, simultaneously overlapping from wet scrubber number one on to wet scrubber number eight, the work could be done in fourteen days. I got Neil's approval and then had a meeting with everyone concerned—gang leaders, foremen and general foreman, including the three potline superintendents. I explained the weather predicament, thus the round-the-clock work schedule. There was nary a complaint, and everything went like clockwork.

Ask an Expert

My second major project was to build a railroad maintenance shop and relocate an existing quarter-mile railroad track to the new shop area. This project taught me another life-changing lesson. No one is anywhere near as smart in each aspect of each trade as the people specializing in that particular trade. And, boy, did I ever learn this the hard way.

I laid out the plan for this supposedly simple project, which involved all the building trades including the railroad track crew. This latter group was to take the existing quarter-mile railway and relocate all the thousands of heavy creosoted ties, plus rails, from the existing location to the new railway repair shop 300 yards away.

I had spoken to the heads of each crew far ahead of time to get their personal input. Then I held a meeting with all concerned and carefully laid out all the details of each step of the process on a three-foot by twelve-foot piece of paper on the office wall. Each crew head agreed with my plan, including Stanley, the gang leader of the railroad track crew, and Cliff, his foreman.

Stanley was a little middle-aged Polish man, weighing no more than 135 pounds soaking wet, but he was very tough and a hard worker. Stanley didn't actually say anything at the meeting, and I suspected he was intimidated, being at this meeting with all these bigger bosses.

Time passed, and the railway repair shop was going up without a hitch, but there was absolutely no progress on the railway track relocation. I contacted Stanley in person and told him that the repair shop was nearly complete and that he was behind schedule. He had nothing to say, just gave me a blank stare and a shrug of the shoulders.

I asked him point-blank when he was going to start, and he said, in broken English, "Meester Gowan, after shop floor laid, I lay tracks."

I was still concerned and asked him if he had made arrangements to get the bedding gravel he would need for the wooden ties to rest on.

He mumbled, "I get to, when it needed."

I told him the schedule called for the gravel to be laid and then levelled weeks ahead of the ties going in.

He said, "Meester Gowan, you no worry. Ties and rails will be in before the building be finished."

At this point, frustrated, I let him be and approached his boss, Cliff. I explained the situation as I saw it: there was no way Stanley and his crew could perform all the work as I had laid it out in the time left.

Cliff casually remarked, "Stanley has worked for me nigh on fifteen years, and he has never let me down once. So don't worry, your bloody railway track will be laid, all nice and tidy, before the big official opening of the new railroad repair shop."

I thought about going higher up the ladder to get results, but with other worries on my mind, I let it go.

Two weeks later, on a Monday morning, I went down to the wharf area to check the progress on the building itself and the railway. The shop floor had been poured, so Stanley now had no excuse. It was time to get the railway track laid and connected.

He didn't have a phone, so I tracked him down on site. He was working with his crew, repairing a siding by the casting building. I tentatively approached him and, among other things, said, "Well, Stanley, the time has come. The repair shop is now finished and waiting for you to lay a quarter mile of track from the coke-calcining plant to the railway building for the big opening."

Stanley just looked at me and said, "Okay." I could get nothing more out of him.

I had another project that I was trying to lay out, so I didn't check back with him for another two days. Then I rode my bicycle down to the repair shop and nearly fell off my bike. The whole quarter mile of track was relocated and connected to the track in the repair shop floor.

I pedalled back up to the connection point, and there was Stanley and his crew, tamping gravel around the ties. How in the name of all that's holy could he have done an estimated—and agreed upon—780 hours of work in two days?

I congratulated Stanley on accomplishing a miracle, but he wasn't impressed. Then I asked him how he did it. He explained in broken English I didn't understand and left it at that.

In the early 1990s, Alcan computerized all of the pot room operations. In preparation for this, one of my jobs was to record the electric and magnetic fields throughout the plant. I spent over three months going all over the plant with my Gauss meter, taking thousands of readings.

MAJOR PROJECT COORDINATION PLANNER

Sorely confused, I thought, "Well, I'll go over and have a cup of coffee with Gordon and the wharf foreman at the Cable Shack. They are close by and don't miss anything going on, and maybe I'll find out something."

I walked in, greeted Gordon, poured a cup of coffee from their everhot coffee pot and sat down. Just then Bert, the wharf foreman, came in, saw me and started to laugh.

Bert was a big strong man who had served as an officer in the German army. He would never disclose his rank or talk about the war, but many years later I found out that he had been very high up in the army. He was not at all the smile-and-laugh type of man. So I asked him, "What's so funny?"

He looked at me and said, "I just came from the railway repair shop and was talking to one of Stanley's men. They sure pulled a joke on you, didn't they?"

"Sorry, Bert," I replied, "but I don't know what you're talking about."

Bert replied, "You would not believe how little Stanley and his fourman crew relocated all of those rails and ties in one day. I've never seen anything like it."

I replied, "Don't keep me in the dark. Tell me how he pulled off this miracle."

Bert carried on, "Apparently the method that he used was something of a standard back in Poland, where he came from, so Stanley thought nothing of it, but it was absolutely brilliant."

It took a great deal to impress Bert, so I was intrigued. "Well, how did he do it?"

Bert didn't answer right away, but he then said the strangest thing. "I didn't know that railway tracks could be bent like spaghetti."

I said, "Bert, come on! Don't keep me in suspense. How did Stanley do a $20,000 job for $3,000?"

Bert replied, "All right, I'll tell you. He disconnected the tracks at the Bunker C tanks, borrowed the D8 Caterpillar from the cokecalcining plant and hooked one very long cable from one of the

138

winches at the rear of the Cat to the head end of the track. He connected another long cable from the other winch to the middle of the long 45-degree curve of the track and just started pulling.

"It was crazy, but the ties started bursting out of their gravel beds and slid across the open gravel field along with the tracks, smooth as apple pie. He kept adjusting the strain on each cable with the Cat's two winches as they slid along. His crew were busy helping by clearing away any blockages along the way. The right-hand curve of the track was now pulled into a straight line, then reverse curved to the left to enter the railway repair shop. All he had to do then was level the track and tamp in the bedding gravel, and in another day they will be finished."

In my summary of this project I made sure that Stanley's efforts were recognized, and he was forever grateful.

I learned a big lesson from this episode, which I have used successfully over and over again. Don't assume you know how a job should be accomplished. Take the time to talk to everyone concerned privately, one on one, and give them an opportunity to express their thoughts on how to do the job. Listen carefully, and only ask leading questions after they have said their piece. Don't push or rush them. You will be amazed at what you find out. Sometimes one man has known the answer to the problem for a long time, but nobody has asked him or listened to him.

This brings up another aspect of problem solving. If you are lucky enough to know everyone in the crew fairly well, ask the most negative, sullen, complaining member first. Listen carefully to his negative statements, as you may be able to turn them into a positive, with the answer to the problem staring you in the face. Sometimes the biggest complainers have the answers without even knowing it. Or they know it, but because they have been laughed at for other suggestions they've made, they no longer bother telling anybody anything unless you ask them—and they like you.

I learned this again in another of my projects.

A Mighty Big Lift

I received a work order from the Maintenance Head Office in Montreal. I was to raise the sunken 600-foot centre section of the south wall of Pot Room 3-A up about 4 1/2 inches to level the building. The work order came with drawings and extremely detailed instructions on how to lift this monstrous wall.

Having been embarrassed by what happened in the railway track project, I was now determined to find out all the background information before I started anything. After all, I hadn't a clue how to go about raising a sunken pot room wall, with all the equipment and other factors that went with it: crane tracks, electrical piping, instrument piping, floors, you name it.

I phoned my old friend Joe, the head surveyor for the plant, and he took me on a field trip to familiarize me with the Column Settlement Problem.

Joe was a very nice fellow, originally from Yugoslavia, with a great sense of humour. He took a great interest in politics and ended up a hard-working city councillor and, eventually, a regional director (with some thanks to yours truly).

On the field trip, Joe told me the plant had been built on a swamp. Each of the fifteen potline buildings were over 1,200 feet long, and they had to be level to operate properly. The civil engineers involved in the original construction took very detailed settlement calculations for each of the forty-eight columns on each side of each building. Using samples, they calculated the settlement of each individual column, and from these calculations they determined the size of the concrete base plate for each column to sit on. Some base plates were four feet by four feet; others were all the way up to eight by eight, depending on the calculated settlement rates for that specific column. According to Joe, this concept and challenge created worldwide interest among civil engineers. Visitors came from everywhere to study their results.

Now that the plant was in operation, Joe's surveyors had a maintenance schedule they followed to regularly monitor and record the settlement in over 1,400 columns. When columns settled an inch or two, major maintenance was notified. Someone went out and loosened the bolts on the cross members to relieve this stress, then they would retighten them. This work took up all the time of two millwrights for the full year.

Next I went to see my friend Mike, the general foreman on potlines 3 to 5, to get information about building settlement from him. He explained that what motivated this particular work order was the sudden, and extreme, settlement of the south side of Pot Room 3-A, which had caused a serious accident with molten metal. Not only had the wall sunk 4 1/2 inches, but all the track and the overhead bridge crane had sunk as well, and nobody had noticed it. They had been moving a crucible full of molten metal from the north side of the room. The crane had just lifted the crucible high up above the pots when suddenly the crucible, crane, bridge wheels and cable started rolling across the deck of the crane, slamming into the south wall. The crucible spilled molten metal at 1,900 degrees Celsius all over the place, nearly killing two men.

I thanked Mike for this helpful information. Then I went over to where the accident took place, trying to visualize who was going to be involved where and with what equipment. I also reread the Montreal engineers' very detailed instructions, which set out everything that would be required.

A team from major maintenance would be the prime people to lift the columns. Relocating the twelve-inch hot steam line and the air pressure lines would naturally be a job for the Pipe Shop. Electrical would take care of all the high-voltage piping on the walls, which went to the anode motors for each pot. The Instrument Shop would rewire the pot control instruments and controls connecting to the potline wall.

The cement crew and the carpenter crew would break out each concrete column and install the rebar on the column platform, forming and pouring the concrete as per the Montreal drawings. We would need the full cooperation of the potlines as, according to Montreal, we would be shutting down pots temporarily, affecting the production of aluminum.

We were also going to need a big load of rolling scaffolding in order to do this job from the outside of the building, as per Montreal drawings. I wondered if the engineers back east had taken into consideration the amount of rain and snow we got. All this column work would take place directly under the eaves of the potline roof, with men working right under the drips or icicles.

Montreal had supplied all the drawings and nearly oversupplied detailed instructions on how to do every step of the job. I got the impression what they had sent showed the process that was "Standard Practice" back east, and I assumed they had done this job many times before in other plants.

This should have left me with no responsibility if anything went wrong. But "once bitten, twice shy." I decided to call a meeting of all concerned. I would go with an open mind, getting their input before I started laying the entire project out. At the meeting I showed all of them the drawings and the very detailed instructions. The meeting went as planned, but because the project was laid out by "Wise Men from the East," there was not a single comment or any voluntary input from any of the Kitimat employees. A big red flag went up in my head.

As I had found out the hard way, to get people's true feelings about a project, I had to go to them one on one, preferably at the project site or in that person's office. I waited a week, giving each of them time to stew over the project. Then I consulted everyone separately at the worksite.

First on the list was Martin, the foreman of major maintenance, the key group. I phoned Martin and asked him to meet me on site. Martin was built like a bear and was very powerful, with an intense stare. You had to handle him with kid gloves if you wanted his cooperation. He was originally from Germany and was proud of what he had made of himself in Canada. He had learned the millwright trade very well and took his job extremely seriously.

We met on site, inside the building, and looked at the worst sunken column with the Montreal drawings open in front of us. Martin mumbled and he grumbled, and then we went outside to look at the columns. The potline walls were covered in aluminum sheeting. He

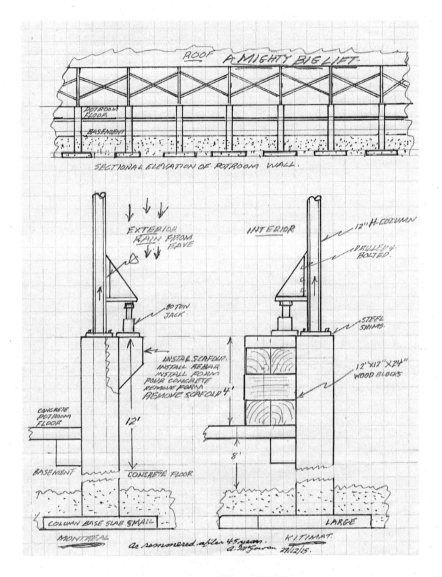

Potline 3-A-South: wall lifting.

grumbled that he would have to cut large holes in the sheeting in over twelve places at each column, then patch the walls back up after the work was done.

He didn't like the idea of jacking the wall back up off a newly poured haunch on the side of the concrete base column. He was also concerned about putting an eighty-ton jack on a newly poured concrete haunch; he was afraid of breaking it off, since concrete takes a long time to reach its full strength.

We came back inside the potline, and he stood staring at this one column for a very long time without saying anything. This was a man who had been adjusting the cross members of these columns for settlement for the past ten years. You could nearly hear the gears grinding in his head.

Finally he looked very intensely at me and said a strange thing: "You do know that I have a large magnetic drill and can very easily drill holes in the upper steel column from the inside of the pot room."

One thing about Martin: he was loath to go against any written instructions, but he was strongly hinting what he would like to do, contrary to the very detailed instructions we had received. Martin didn't seem to want a reply. But I knew that he didn't like what Montreal had laid out, and some alternative idea was slowly forming in his head. Then again, knowing Martin, maybe he was deviously planting the seeds for the alternative idea in my head. Time would tell.

Some supervisors seem to have two personalities—one when they are out of their office, and the other when they are safe in their office. The office and its reinforcing surroundings make them feel more secure and expansive. I decided to take advantage of this.

I waited a few days and then casually dropped into Martin's office, taking the blueprints with me. I took the bit in my mouth and without any preamble asked him, "Well, what do you think of the jacking-up process?"

He looked at me with those intense eyes of his and finally said, "It doesn't make sense, doing all this work outside on scaffolding in the fall rain underneath the eaves, when we could just do it from the inside, nice and easy at floor level." This was a big freethinking breakthrough for Martin. But after all, he had been bracing columns very efficiently his own way for more than a decade, lifting columns an inch at a time by slipping one-inch metal pads between the concrete pier and the large steel base. Why change to a costlier, less efficient way?

Happy with Martin's unconventional method, I approached the instrument foreman, the electrical foreman and the Pipe Shop foreman to ask about the methods the Montreal engineers wanted us to use, which involved splicing five inches of wire or pipe into all the wiring and air pressure pipes serving the fifteen columns we needed to raise.

The four of us went out and looked at all the piping going from the pot over to the wall, where it was fastened solidly. Tony from electrical suggested that all we had to do was loosen the brackets holding the piping on the walls, letting the piping slide through while the wall was being raised. He figured that all the piping and conduit should slide through the brackets. Any that didn't would be repaired. Then they would refasten the brackets.

Committing them, I asked the burning question, "Do you all agree that this is the way it should be done?"

They all verbally agreed. Two problems solved. One more to go.

I repeated that the specifications from Montreal called for the pot's instruments and 440-volt power to be turned off while the walls were being lifted. Nearly in unison the three foremen said, "That's bullshit. We will repair any shorts, if there are any."

I now focused on Eric, the Pipe Shop foreman, and said, "While we're out here, let's take a look at that twelve-inch high-pressure steam line running north, high above the centre passageway. They have a detailed procedure for bringing it back to its original position."

We walked over and, standing below the pipe, shone our powerful spotlights 40 feet up to look at this steam line.

I said to Eric, "Montreal calls for this steam line to be shut off, and all hangers loosened off, before we raise the wall."

Eric exclaimed, "You can't do that! Shut off this steam heating for all the offices, plus the men's showers, lockers and lunchrooms, for three weeks? We would have a rebellion!"

I explained, "From their point of view, they are concerned that moving the pipe could break it, emitting high-pressure steam."

Eric just laughed and said, "Crazy as it might seem, the opposite is the truth. You can move or bend hot high-pressure steam lines easier then you can when they are cold. We don't even need to shut the steam off. We'll just loosen all the hangers. It's only going to move, at the most, five inches over a 61-foot span. Plus a hot pipe will bend easily, like spaghetti."

I asked him if he was happy with this method, and he said yes.

The entire job progressed smoothly over the next two weeks without any major problems. There wasn't one instrument, electrical conduit or airline damaged, including the big twelve-inch hot steam line. Then everything was fastened back in place.

One large flaw. I hadn't thought to get engineering to certify all of these changes from Montreal engineering's specific instructions.

Now comes the inexplicable kicker. I thought we had done the very best job that could be done, under the circumstances. Montreal had estimated the cost at $120,000. By thinking outside the box, we had done it for $38,000. I thought that we would all be congratulated for doing a good job and for saving a pile of money to boot. No way, Jose! I now learned another lesson about large organizations. The mid-level echelons were not happy, basically because we had made them look incompetent, especially Montreal engineering.

In retrospect, if I had to do it over again, I would have intimately involved Montreal in each decision.

Nothing was said to me directly, but I could feel the chill. Months later I got a left-handed, flippant compliment from my superintendent. But nothing in my personnel file, good or bad. It is no wonder to me that some people develop a negative attitude when it seems you just can't win.

Nevertheless, I was very satisfied with the team effort, and with my new tactic of forcing "How" into the equation. We had saved a fortune and established a better standard for doing this job by allowing the workmen to do it from the inside the building. This also reduced the safety concerns.

I asserted myself and independently sent thank-you notes to all concerned for their input and for "thinking outside the box." That note went over like a lead balloon. I gained the animosity of upper management, and the respect and approval of the people I had to work with every day. The weird wonders of bureaucracy. Sometimes you can't win for losing.

The Double Switcheroo

Five years after we started buying rental units in Terrace, Mary and I decided that in order to look after them properly, we should live in Terrace. I went to Jerome, my boss in maintenance, and explained the situation to him, then asked for official permission to move to Terrace and commute along with over 300 others from the plant. He checked it out up the line before letting me know that it was officially approved, but I could see that he didn't like it.

I continued on as major project coordinator, and a couple of months later we bought a house and moved to Terrace, to much complaining from our three children, two of whom were teenagers. Since my job sometimes required me to work overtime, I bought a car and kept it in the inside plant parking lot. This gave me a means to get home to Terrace long after the bus left.

Around the same time, Canada was sliding into a huge recession, and a notice came out that Alcan was going to cut the salaried staff at Kitimat by 60 bodies, out of 360. All my projects were being cut back, and I was parked way out on a siding, figuratively speaking, away from the main line. Management had a tough job, cutting the staff by sixty within three months, and one day a friend came into my office and said that he had heard I was on the list to be laid off from the maintenance department.

A day later I got a phone call from my superintendent, Dean, to the effect that he had heard there was a rumour going around that I was to be laid off. He wanted to assure me that this rumour was not true.

Strangely, this reassurance didn't make me feel any better. Old-time friends were leaving right and left, and being laid off in a small company town was a major calamity. With no local jobs available, you were forced to move, taking a beating when you sold your home.

I was living in suspense until a week later, when my boss came and told me that my major project coordinator job was over, and I was being moved to pot room maintenance, as their planner. Apparently Tom, a new employee and the current planner at pot room maintenance, was not fitting in, so he was being laid off instead of me. It was a great relief for me, but not so good for poor Tom.

I closed down all my cancelled projects and filed them, then moved over to pot room maintenance.

I had now been working for Alcan for seventeen years, in three different positions, and I was completely overwhelmed by the considerate, caring atmosphere in my new department. Everybody worked together pleasantly, and there was no tension in the air. I fit right in.

It was a full year later before I discovered the truth regarding the rumour that I was to be laid off.

Ole, my foreman, took me out for lunch one day and brought up the subject of the large staff layoff. What he said shocked the hell out of me.

He started by saying, "Alan, regardless of what you were told, you were on the list to be laid off. Would you like to know what happened?"

I replied, "I certainly would."

Ole continued, "Bill, our general foreman, went to a maintenance meeting to discuss all the layoffs, and when your name came up, Bill was upset. Apparently the job you did in raising Pot Room 3-A impressed him very much, and he felt you could improve pot room maintenance if given the chance. He felt you were too good to be laid off, and he told his boss to lay off Tom instead."

I replied, "I take it that you have told me this in confidence, and I appreciate it."

I sure looked at Bill differently from that day forward. He was one of the best bosses I ever worked for, a real gentleman in the true sense of the word.

Apparently the main reason that I was on the list was they didn't want any members of management living so far away from the plant. But Bill's decision was respected, and I was protected.

CHAPTER SEVEN

The Early Riser Cooperative Bus Line

Setting up the Bus Service *151*
Ice, Snow and Rain *156*
Don't Stop *158*
Too Close a Call *159*
Snow Blind *160*
Buying a New Bus *161*

Alan and Mary in new Terrace house, 1972.

Our house in Terrace in 1975.

We made the big move to Terrace in the fall of 1970 to be near our rental properties, and I settled down to a commuting life. I joined a carpool, which was an eye-opener. As a part-time mechanic I couldn't believe how careless people can be with the vehicles they drive.

The four of us took turns, each driving for one week. Tom ran his station wagon in winter with no weight in the trunk and no snow tires. Dick bought two to three gallons of gas every morning, hoping to make it home without running out of gas. He never explained why he followed this stupid practice. Harry drove like a maniac, ending up in the ditch twice the first year, and completely turning the car around on a straight stretch of road at 90 clicks.

Ron was commuting with another group, and we exchanged stories. As a result of that, we decided to set up a meeting for everybody commuting to Kitimat. Forty-five people attended, each with something to say about commuting. One man had done a study and found there were over 450 men and woman commuting the 50 miles. Another had done research after he broke his back in a commuting accident and found there were numerous workers who had serious accidents on the road. There were also complaints about the snow clearing in the winter—or lack of snow clearing, with tales of cars trying to get through three feet of snow that hadn't been cleared for three days.

The upshot of the meeting was that we were highly motivated to get some kind of bus service going between the two towns.

Setting up the Bus Service

When I was on holidays during my first summer in Terrace, I did some research on various cooperative bus services, one in Surrey and one running from Castlegar to Trail, and learned a great deal about how to set something up and how to avoid the pitfalls.

We got through the first two winters, still driving cars, when Noah, the owner of the local intercity bus line, approached Ron and me with a proposition. He would supply a bus if we would organize the times and the bus stops. He would charge us seven dollars per day. We thought this was great.

Two weeks later we had a school bus and a driver, but only for dayshift, and it was like pulling teeth to get people to break up their carpools. They were a semi-social thing; in the summer, some would stop every day and pick up a case of beer to drink on the way home. After two months of advertising, talking and pleading, we had broken up four carpools and had a total of fourteen passengers.

At the end of three months, though, I could see that the expense of paying a driver who brought an empty bus to Terrace to pick us up in the morning, and then returned to Kitimat with an empty bus at night, would never work. I was half-heartedly looking around for an alternative if they decided to cancel the service.

Sure enough, we were going home to Terrace on a Thursday afternoon in June when Noah got on the bus at the smelter site and rode with us up to the Kitimat townsite. Without discussing anything with Ron and me, who had been helping get riders for the service, he stopped the bus and stood up to address the fourteen of us. "This being Thursday," he said, "as of Monday this service is going to be cancelled." Then he got off the bus at the city centre, case closed.

His attitude really made me mad, so I got up and made a foolhardy statement: "As of Monday, come hell or high water, I will have a bus on the road."

After sitting down I started to realize the awkward situation I was in. If a bus stopped running for any amount of time, everybody would revert to carpooling and we would never get them back again.

Walking home from the bus stop, my mind was in a whirl. How was I going to get out of this mess? And if I was able to keep it together and get it off the ground, boy, was I going to be busy. I discussed it with Mary, and she put forward various solutions from her more objective position.

I needed a driver with a class one licence, and a bus, by Monday morning. So first thing I did was phone up Albert, who worked for

Alcan and had a class one licence. He was in a carpool, but he said that if he got paid, he would drive the bus starting on Monday, no sweat.

Now all we needed was a bus. I chased down Hank, the owner of all the local school buses in Terrace, at the Legion and when I asked him if he had any buses for sale. He replied, "As a matter of fact, I have a spare bus that is in pretty good condition. You can have it for $3,500."

I took Friday off and met him at his bus depot. The ten-year-old bus looked a little grungy, but it didn't burn any oil and the tires were very good. I told him I would buy the bus if he would teach me how to drive it. He spent the better part of the day teaching me how to "double clutch," "split shift fourteen gears" and basically drive a big school bus. It didn't have air brakes, so I would not need an air brake certificate, but I would need to upgrade my class four driver's licence to a class one. I spent the whole of Sunday driving the bus around a vacant parking lot.

I paid for the bus out of my own pocket, transferred the title and drove it over to Albert's so he could drive it to work on Monday morning. But, surprise! He had been transferred to work at Kemano power station for three months and wouldn't be home.

Guess who drove the bus for the next three weeks, without a licence? Little old determined me. I got my licence during the day in Kitimat—thankfully the examiner never asked me how the bus got to Kitimat—and I paid for the special insurance we needed for a public bus. By now my wallet was getting pretty thin.

I charged each passenger $100 to join the newly incorporated but not yet registered Early Riser Cooperative Bus Line, and five dollars per day to ride in it. The first week we had 10 membership riders. The second week 15, the third 20. It took a year to get it up to 24 passengers, which meant it was paying for itself and starting to pay me back all the money I had subsidized it with, though nobody seemed aware of how much I had gone out of my way to get this bus running. Especially Steve (My Cross to Bear).

One of the first things I did was appoint a conductor, and here I had a big stroke of good luck. I appointed Dirk, who was Dutch. I don't know how I could have been so lucky. I used to sit there in amazement as Dirk did his thing. He was absolutely merciless. If people didn't or couldn't pay, they were off the bus, no excuses. And you never heard so many excuses in your life. Here are a few typical excuses, to which the conductor's standard reply was "Pay or off the bus."

a. Dirk, I'll pay tomorrow.
b. I'll pay tomorrow. Don't you trust me?
c. I'm broke, but I'll pay on the way home.
d. I'm getting paid tomorrow.
e. I am a director of this bus cooperative. Don't you trust me?

Finally all of these members fell into line.

One day I was over at Dirk's place, going over our books, when I asked him, "Dirk, where did you ever learn to be so ruthless about collecting money?"

His wife spoke up and said, "You would have to grow up in Holland to learn how."

I wasn't sure what she meant, so she told me, "When we were growing up, everything was paid for by the week. People were coming to your house every day for weekly payments of rent, heat, power, papers and milk, plus ice and vegetables, and there were no excuses. You paid or you were cut off."

Take note: if you want money collected, hire a Dutchman and you won't have any worries. Dirk lasted eleven years, until he retired. Then I hired another Dutchman, and he was as good as Dirk.

An interesting episode happened after we had been operating for over a year. According to our Early Riser Cooperative Bus Line charter, and our insurance, we were not allowed to take anybody who wasn't a member of the co-op. One morning, when I wasn't driving, a stranger got on the bus and gave Dirk, the conductor, a long-winded story about wanting to get to Kitimat, and could he ride the bus for five dollars. His patter seemed too slick so I spoke up and told inflexible Dirk to let him on the bus for free. Then I asked the fellow to come and sit with me, and I grilled him.

My first bus, 1971. Bought for $3,200 and sold $100 shares to 34 and started the Early Riser Bus Line, Terrace to Kitimat.

Our first bus and the board of directors of the Early Riser Cooperative Bus Line, 1972.

He couldn't tell me any details of who he was going to see, or who had hired him and for what. He got off the bus at the city centre instead of at the plant. When I got to my office, I wrote all the details down and got Conductor Dirk to witness it.

Two weeks later we received a letter from ICBC, notifying us that we were being accused of allowing anybody that paid to ride our bus, contrary to the co-op regulations. An inspector of co-ops was going to be at our next board meeting, and we were in danger of losing our charter.

Who was trying to wreck our co-op bus line?

I had a few political friends in Victoria, so I got on the phone and found out who was behind all this kerfuffle. Turns out Noah from the intercity bus line had spent the money and hired a private investigator to fly up all the way from Vancouver and try to ride our bus as a private citizen, just so Noah could lay a charge to get rid of the so-called competition.

The inspector came to our next meeting and quickly realized that he had been set up. He dismissed the charges.

Operating a workers' bus had its good and bad moments. We developed a set pattern: two riders who were born late were always accusing us of leaving early. I bought a very large, brightly lit digital clock and installed it above the dashboard for all to see, and each morning a countdown was chanted in unison for the last twenty seconds before departure.

In the morning, after we got to the outskirts of town and made our last pickup, we started moving. Then God help the loudmouth who was still talking as everybody but the driver went to sleep.

Ice, Snow and Rain

Driving that bus back and forth to Kitimat in the winter, we had one major problem. Halfway to Kitimat was a two-mile-long ridge that was 250 feet higher than the land on either side. If we got six inches of snow in Terrace, you could nearly guarantee that there would be twenty-four inches on this ridge. It wasn't cleared because the highway people didn't know about it, and sometimes it would be a couple of days before they sent out a plough to clear that stretch of road.

Highway survivable, says veteran

The Terrace-Kitimat highway has been very much in the news recently. There's one Works employee who's driven it for 15 years, and he says it's survivable.

It's a story that seems to occur much too often. With shocking frequency, we're horrified to learn that Highway 37 has claimed another friend or relative.

Does the fault lie with the highway or the drivers?

We put that question to Al McGowan of reduction development, who with several others, formed the Early Riser Co-op Bus Line in 1971. Al estimates he's driven some 360,000 miles since then . . . and he's never had an accident, not even a minor fender-bender.

"If I were minister of highways, I'd place a cross at the site of every fatality on that road."

But on a Friday afternoon — Winter or Summer — especially before a long weekend, look out! You never know what they'll do . . . and that's the scariest thing you have to face."

Today, there are 106 Works employees living in Terrace and vicinity. Al says that at one time that number was 356. Of the 106, 35 are members of the Co-op, which rotates the driving chores among four men with Class II air licenses including Al.

No interest in another

He says that for years, he's been trying to organize another bus system for shift workers, but there has been relatively little interest.

"Car pools are very social things . . . people become very used to them. But many of the people who joined the Co-op have been in accidents on that road and just won't drive it any more."

And not all the car pools are havens of security, says Al.

Al McGowan sits at the wheel of the Early Rise Co-op bus, which he reckons he's driven 360,000 miles . . . about 15 times around the Earth. He says he feels very comfortable there, despite the odds which show a Terrace-Kitimat commuter can expect to be in an accident once every five years.

that every driver signalling a right turn will turn left; also, assist every vehicle that wants to pass you or he'll involve you in his accident; count on your engine to pull you out of a tight situation and it won't; always give every other driver the benefit of the doubt . . . and the right of way; and finally, underdrive every situation and you'll survive another day."

Terrace Herald article.

It got to the point that when I arrived in Kitimat, I would phone the highway maintenance operator about the conditions. But, strangely, nothing was done. We tolerated this weird situation for over a year because I couldn't think of how to motivate them to send a snowplough out to clear two miles of road in the middle of nowhere.

Finally, the light went on. I figured out how I could shame them into clearing the snow. I called the Kitimat CBC radio station and set up an appointment for 8 every morning to give a report on the road out to Terrace.

This went over very well. Everybody in Kitimat looked forward to my report and waited for it before venturing out to travel to Terrace for shopping or the airport. Some even phoned in to thank me. Through the back door we got a snow station at the top of the ridge that was staffed every night during the winter to keep the road open.

Sometimes you have to play dirty to get things done.

If I was driving the bus, I would wave at the poor snowplough driver, just sitting there, all alone every night. Desperation sometimes forces you to think outside the box.

Don't Stop

I fired up the bus one normal winter morning with six inches of fresh, very wet, heavy snow everywhere. We did our three main pickups and then headed up the 300-foot airport hill with a full busload of workers. As I drove I noticed that there hadn't been any snowploughs out yet, so we were down to 35 miles per hour. On the upper airport flat I managed to get the bus up to 50 miles an hour, but even with the weight of all the passengers, the bus was still hydroplaning in the heavy, wet, sloppy snow.

It was scary, to say the least. I could turn the steering wheel 30 degrees and absolutely nothing would happen. But if I turned it another 5 degrees, the front tires would suddenly catch, and if I wasn't quick enough, we would be in the ditch.

Speaking of ditches, we had already passed three abandoned cars, parked way off the road.

I decided that travelling at 35 miles per hour was the safest speed. This meant aggressive drivers were starting to back up behind us, beeping their horns and flashing their lights, wanting to pass us. I guess they didn't realize that I was breaking trail for them, as it was untouched snow ahead of me.

We now had a tense situation both inside and outside the bus.

When I started pulling over to offer help to the first occupied car that was in the ditch, the whole bus started yelling, "DON'T STOP!" Someone said, "To hell with them. They were going too fast, so it's their own stupid fault they're in the ditch." There were other nastier remarks, and some comments to the effect that if we stopped for all of them, we would never get to work.

A car pulled out of the line behind us and attempted to pass us on the six inches of undisturbed snow. Frantically I pulled over to avoid a sideswipe. He shot by me, doing at least 50 miles per hour, tried to pull back in in front of me, promptly started fishtailing and literally flew off the highway into the ditch.

Going down the long hill to the Lakelse Flats, nobody attempted to pass us because of the huge drop-off on the left side. Once we got down to the swamp flat, though, everybody wanted to pass, and away every one of them went, flying into the boondocks.

In desperation I turned on the hazard flashers, but that didn't work, so more ended up in the ditches. At this point I got a bright idea that also ended up not working. I asked one of the guys to go to the back of the bus with a flashlight and signal the following cars not to pass us. The car driver understood it to have the opposite meaning, and another one hit the ditch.

What to do, what to do? Finally I pulled over, off the road, at the Hot Springs Garage and let the whole stream of cars go by. I asked everybody if they had any ideas about what we could do. All I got was variations on "If they are stupid enough to try to pass us and go in the ditch when we are breaking a good trail, too bad. I want to get to work."

Back on the road again, ploughing snow, with cars still passing us and hitting the ditch, we ended up forty minutes late for work. We also passed thirty-four cars in the ditch. Somebody on the bus took the time to count them.

Too Close a Call

I was driving the bus to work one fall day in 1975, through heavy rain that had been pounding down for three days. It was pitch-black, misty and miserable. I had a hard time seeing clearly, so I was going slower than usual. Looking in that large rearview mirror I could tell that all the guys were nervous and not sleeping. We came down the Onion Lake Hill and then crossed the flat, where we could see the engorged river running alongside the highway.

We went around a sharp corner with the river to our right. The wipers were going as fast as they could, but in the darkness and rain I could hardly see the black pavement in front of me.

Something, or someone, made me stop. I got out and walked along the road. No more than 20 feet in front of the bus the road was completely gone. In its place was a 10-foot drop into a raging hole of water,

50 feet long and god knows how deep. If we had gone another 30 feet, every one of us would have been drowned in that hole.

Luckily we were the first ones there—at least, I hoped we were—and between us we and the next few drivers put up flashing lights to warn of the hazard. One fellow who came along behind us volunteered to stay there, with his car, while we drove on to Kitimat. We were able to go off on a side road, an old logging trail, to get around the washed-out section of the highway. Once we were at work, I phoned the highways department to come out and put up barriers, but it took them nearly two weeks to backfill the large hole after the river went down.

Snow Blind

On a beautiful, cold, late-winter day, the sun was shining between sudden gusts of snow, and we were heading home on the bus. Two feet of snow had fallen while we were at work, over and above the eight feet already on the ground. The conditions were as different as possible from those we encountered when the road washed out a few months earlier.

Away we went, past the town, past the dump. We were on a straight stretch when suddenly I couldn't see a damn thing outside the bus—just snow and more snow. To make matters worse, the sun was out, shining through the snow coming down. I had to pull over and stop because I couldn't see the road or the six-foot snowbanks on either side of the road. Just bloody white all around us. I was completely dumbfounded.

I sat there in the driver's seat for a minute, trying to figure out what to do, and finally turned around and said, "Boys, I can't see a bloody thing. We will have to stay here until dark. Maybe with the lights on I'll be able to see to drive. Have any of you guys any other idea of what we can do?"

One bright shining soul suggested cutting down two 15-foot poles and sticking one out each side of the bus. When one of the poles touched a snowbank, he would yell for me to turn either right or left. In other words, we were going to attempt to drive blind. To make matters worse, there wasn't another vehicle breaking trail either way, which was very strange.

Away these guys went with an axe. They cut two long poles and peeled them, then lugged them into the bus and shoved them out the windows. We started off again, very very slowly, with much confusion. We must have looked foolish, going down the highway with these long poles sticking out the windows. And what would have happened if we had met a truck speeding either way? The impact would have smashed a window, and what about the poor schmuck who was holding the pole?

We hadn't gone half a mile when a new fellow on the bus yelled, "STOP!"

He came forward to me and said, "I'm new on the bus, and I didn't want to interfere, but when this method became dangerous I had to. Throw those stupid poles away. I have a better solution. I used to work up north on the DEW Line radar stations, and I've had experience with snow blindness. Have you got a pair of sunglasses and some masking tape?"

I found both items in the glove compartment and gave them to him. He cut off four small pieces of tape and stuck them to the lenses, leaving a one-eighth-inch slit in the middle of each lens. He gave them back to me and said, "Put these on and let's get on home."

I put these taped-up sunglasses on and, amazingly, I could see perfectly. I could hardly believe it, and away we went. We all had learned something.

Buying a New Bus

Our old second-hand bus had 60,000 miles on it when we bought it. After eight years it had 140,000 miles and was definitely showing signs of old age, so the directors agreed that we needed a new bus. We had built up a contingency fund of $30,000.

We set up a committee to define our needs, then went out shopping for a school bus with our added custom features of high-back seats, double-pane windows, eight heaters and a large gas engine with overdrive. The price came to $72,000, but we had only $20,000 that we could use, so we needed to borrow $52,000.

I shopped around at the credit union and all the banks, but the lowest interest rate I could get at the time, on what was basically an unsecured loan, was 9 1/2 percent. My reception by all the loan officers and bank managers was rather frosty, and they were clearly not interested in my proposal. When they got out their sharp pencil, they could see that with only thirty-four riders per day we couldn't afford the payments.

I approached Alcan's local personnel manager and explained who we were and what we were accomplishing: supplying safe, relaxed transportation for workers commuting back and forth from Terrace. Then I asked if there was any chance we could benefit from Alcan's position and get the company to co-sign a loan with us so we could get a better rate.

From the look on his face when I had finished my presentation, I realized that I was asking something rather unusual. But he simply said, "Leave this with me and I will get back to you within a couple of weeks."

A week later I received a phone call from Mr. Personnel Manager: "Go see the manager of the Bank of Montreal in Terrace when you get back tonight at 5 p.m. He will be waiting for you."

This was very direct and to the point. The only problem was that the bank closed at 4 p.m., and based on the attitude the manager had shown before, I couldn't see him keeping the bank open for an hour for little old me.

After dropping all the passengers at the various stops, I parked the bus and went over to the bank. It was all closed up, but when I went to the large glass front door, there was the manager standing on the other side. Surely he wasn't standing there watching for me.

But he was. He opened the door and ushered me into his office as if I was somebody important. We sat down in his office, and after serving me a nice cup of coffee, he presented me with an application for a three-year loan at 6 1/2 percent! I couldn't believe it. They had

Our new bus, 1975.

knocked 3 percent off. I instantly signed the paperwork before he had a chance to change his mind.

I'd heard of "preferred customers," but I just thought they were characters in fairy tales.

I hastily called a meeting of my board and they voted in favour of buying the new bus. Three months later we had a school-type bus, designed for long trips, and we ended up putting 1,296,000 miles on it over the next sixteen years.

Being of a curious nature, and also being naïve about large banks and companies, I was interested in finding out exactly how a bank could lower an interest rate, and at whose discretion. As I started my investigation of banks, I got a breakdown of who was on the Alcan board of directors and then started looking at what other boards they were on.

Bingo!

You guessed it. One of the Alcan directors was on the board of the Bank of Montreal. All it took was a simple phone call from the Alcan director to the director of the Bank of Montreal—or that's what I assume happened. But being able to drop the interest rate by 3 percent? That really impressed me. What power!

I continued studying each of the Alcan directors and was fascinated to discover the company's brilliance at picking directors who also sat on the boards of other key companies that Alcan dealt with every day, like transportation and finance companies, to name two.

I served as president and chief driver for twenty of the twenty-two years the bus ran, and I drove around the world seventeen and a half times.

CHAPTER EIGHT

Unusual Incidents and Encounters

Put It Back Up *167*
Catch the Rats *168*
Gassed *170*
An Open Secret *171*
Day with Dad *173*
Wildcat Strike *176*
Wild Goose *179*

Joe and Alan back from fishing, 1960s.

By the 1970s, the smelter in Kitimat had been in operation for two decades. Things had settled into a routine, but I don't know if there was ever a regular workday. At least not for me as the pot room maintenance planner.

Put It Back Up

It was late winter 1972, and we had an unusual northern snowstorm, very cold with a fairly heavy wind that would blow the landed snow wherever it liked, building up large snowdrifts with the five or more feet of snow that had fallen over the previous twenty-four hours.

I got to the pot room office as usual at 7:30 a.m. in spite of this storm. Ole was already behind his desk, but as I approached my office he got up and came over to talk to me. He had never done this before, and judging by the expression on his face, I knew we had a large problem. He didn't even respond to my salutation with a "Good morning," but said, "Before you do anything, go out to line 1 and 2 and take a good look at all the outside exhaust ducting." (Note that each potline was 1,400 feet long, with the exhaust ducting hanging 12 feet off the ground on the outside of both sides of each building. That's 16,800 feet of ducting, 12 to 18 inches in diameter, including over three miles of exterior ducting.)

Curious I asked, "What's going on?"

Ole replied, "Just go and have a look. Then we will try to figure out what to do."

Without further ado I put on my hardhat, goggles and mask and headed out to check the exhaust ducting. I had never seen Ole so serious, so I really wondered what had happened.

It didn't take me long to find out. The ducting was down on the south side of all six buildings, due to six feet of snow that had piled up on that side. This would mean repairing miles of broken ducts and lifting nearly 8,000 feet of ducting back in place, ideally by tomorrow!

Oh man! Did we have a major job on our hands, putting this ducting back up in the dead of winter. We would immediately have to shut off the lines leading to the fallen exhaust ducts so we could get suction in the remaining half of the line. No suction at the pots causes the burners to plug up, affecting production. It also fills the pot rooms with smoke, and the pot room people will not work in the lines if they are full of smoke. So this was one enormous, high-priority job.

We decided we could spare eight of our men to do some of the work. The plant could clear the snow and supply rolling scaffold, but the major work would have to be contracted out, at cost plus, to Leo, a local contractor who had the men and expertise to handle the job. Leo was a big boisterous Italian who loved nothing better than a tough job, especially at cost plus.

Greg and Ole took this decision to a meeting that afternoon with the potline supervisors and maintenance superintendent, who had to authorize it all—with the potlines understanding that they would have to put up with a bit of smoke for at least a month. In six weeks the job was done, and operations were all back to normal. I thought we would at least get a letter of thanks, but no thanks were given.

These freaky snow and wind conditions only happened the one time in my thirty-six years at the plant, but it was enough to curl your hair.

Catch the Rats

One day in our office, upstairs, just off the pot rooms, my big boss came to me and said, "Alan, I understand you are a man of many talents."

Now hold it right there! Any time somebody blatantly flattered me I was immediately on my guard. I knew something was coming down the pike, and it usually wasn't good.

He continued, "The plant has a problem, a rat infestation. Lily [our secretary—a tall, thin, older woman who was always nervous about everything] came to me about it, as she is scared right out of her skin that she might see one in our offices. I'm appointing you to be our official 'Rat Control Expert.'"

It was weird that he had picked me. I hadn't told anyone, but I had just killed about fifty black rats at the garbage dump with Brandy, my Cairn terrier. I had gone to the dump at night with five gallons of fish guts, and when I threw them on the garbage pile, all hell broke loose. Rats immediately came from everywhere, with some even trying to crawl up my legs.

I suddenly remembered that Cairn terriers were bred in Scotland to be "ratters." Brandy was in the truck cab, so I opened the door. She was out in a flash and went completely crazy. She would grab a rat by the neck, shake it twice to break its neck, drop it and grab another one, all in the space of less than ten seconds.

Those rats must've been very hungry, because they wouldn't give up all those fish guts. Brandy had a grand time. When I saw she was finally tiring, I picked her up and put her back in the truck.

Now that I'd been given my new assignment at work, I went to see "Supercop" in Security to find out what was going on with this so-called rat infestation. He said, "Apparently we had an old tramp steamer come into our dock with a load of something a couple of months ago. They forgot to put the rat shields up on the tie-down ropes, and for some reason a load of rats crawled down the ropes and left the ship. They're the dark brown Norwegian rats with hairless tails, and they've multiplied and are all over the plant."

Obviously the problem was real, so I borrowed a ladder from the janitors and waited for Lily to go home for lunch. The trap door to the attic in the suspended ceiling was right above her desk, and I didn't want her around when I climbed up into the attic to check for signs of rats.

Sure enough, when I got into the attic I confirmed we had an invasion of rats. I don't know what they were eating, but there were feces all over and at least two nests. I came back down, closed the trap door and hid the ladder. I didn't want to scare Lily any more than she normally was.

That night I dug out three rat traps, some bait and small chains. If you don't chain your trap down, the caught rat sometimes will walk away with the trap.

I always got into the office at least half an hour before Lily and the others in the morning, which gave me time to crawl up into the attic and bait, set and chain down the traps. Then I closed the trap door and put the ladder away before she got in.

The next morning I arrived extra early—and was I in for a surprise. I had caught a rat! And it was hanging from the ceiling, suspended three feet above Lily's desk!

Luckily no one else was in the office yet, but if this had happened while Lily was sitting at her desk, we would have needed to find a new secretary.

I wish I'd taken a picture of the thing. But all I could think of at the time was getting rid of the rat before anyone else arrived. When I went up to get the rat, I took a look at the situation and decided I hadn't closed the trap door properly the previous day. When the rat was caught, it must have crawled over to the door, flipped it up and fallen through.

Needless to say I straightened everything out, and over the next couple of weeks I trapped a few more rats. But I never said anything to anybody about what I found that morning, just in case poor Lily got wind of what had happened.

Gassed

One nice sunny day I went out for an early lunch on my own. As I walked back to the office through the parking lot, I noticed a forklift idling close to the main entrance door. When I walked through the maintenance office, there was the secretary, Margie, idly chatting with the forklift operator. I guessed that he had brought something and needed her to sign the receipt.

I climbed the stairs to our twelve-man office. When I opened the door, I was greeted by a strange smell, like diesel. Because I had been a painter, and had painted dozens of cars, my smeller was just about burned out. But I could smell this.

Then I noticed that Lily, our secretary, was fast asleep with her head on the desk.

Something weird was going on.

I quickly went around to the other offices. There was Serge, also asleep, and Jerry, all the way down the hallway, was asleep too.

Suddenly it dawned on me. The vertical exhaust pipe of that idling forklift was right under the intake air vent that stuck out of the side of the building. It was supposed to suck fresh air into our upstairs office, but right now it was sucking in deadly gas fumes.

I ran back down the stairs to the maintenance office. Sure enough, there was the forklift operator, still sweet-talking Margie, completely oblivious to the fact he had nearly killed three people.

I told him in no uncertain terms to quickly move that etc. etc. forklift. I then told Margie what was going on upstairs and asked her to get the ambulance here right now. I rushed back upstairs and opened all the doors and windows I could find. Slowly everybody woke up, and surprisingly there were no aftereffects.

When everything had calmed down, I suggested to my boss that I would make up the drawing and work order to have a steel guard installed under the vent so this would never happen again.

An Open Secret

I was in my fishbowl office in the Pot Room Maintenance Office, with Ole to my right, Russ to my left, and windows all around.

It was 9:30 a.m., and everybody was working away when I caught an odd movement to my right. Emil, an electrical foreman, also known as "Dis over Der," had stealthily entered Ole's office, looking furtively all around. What was up?

He was carrying, with difficulty, a very large cardboard box. Emil was a short, heavyset man in his middle forties who had come from central Europe. He was an electrical foreman and good at his job, but very excitable. You could always tell when Emil was getting excited because he would revert to basic English, either starting or finishing every sentence with "Dis over Der."

Now he was glancing around as he tried to surreptitiously slide the large box across the floor, right by the corner of Ole's desk where I couldn't see it. Then he spoke quietly to Ole, and Ole looked around. We had a pass-through between all our offices, but with the steady hum of the pot room fans around us, I couldn't quite hear what they were saying.

Being curious, I got up and went into Ole's office—it was nearly coffee time anyway—and caught Ole trying to hide a beautiful, expensive, six-volt spotlight. I asked both of them, "Hey, where did you get such a nice spotlight?"

Emil stammered and said, "It's a secret, Dis over Der. Do you want one?"

I said sure, and he opened up the big box and gave me one.

I thanked him, and after admiring it I put it on the window ledge.

Emil looked horrified and said, "Dis over Der, don't show it to everybody. Put it away."

I could have asked him why, but I thought I could get a better answer from one of the other electrical foremen.

Russ, the other maintenance foreman, came in from his office for ten o'clock coffee and saw the spotlight. Russ being Russ, he had to know what was going on.

Ole just said, "Russ, if you shut up for once, Emil will give you one."

Russ said, "Sure, I can keep my mouth shut."

Ole then dug one out of the big box and gave it to him.

Emil then said, "Dis over Der, gotta go," and quickly left the office.

Greg, our general foreman, came out of his back office for coffee. When he spotted all the lights, he was most curious to know who had paid for them. Ole told him that the electrical department must have paid for them, adding, "There's one here for you," before pulling one out of the box and giving it to him.

I was still curious. "Emil has never, ever given anything to anybody," I said. "He's as tight as a shark's ass, and that's watertight. Why would he pick us to give twelve spotlights away? Something doesn't add up?"

The next day I was passing the electrical office and spotted Peter, another electrical foreman, and thought I would drop in. He looked up from his desk and said, "I'll bet you I know why you are here. You want to know where Emil got all those spotlights, don't you?"

I said, "Among other things, yes, I was curious, especially coming from old Dis over Der."

Peter said, "Well it's really very simple. Emil's crew needed some spotlights to check out the potline overhead cranes. He thought he filled out an order for six, but he mistakenly ordered a gross because he thought one gross was six and didn't know it was a hundred and forty-four. The requisition surprisingly got authorized all the way up the chain, including the buyer.

"One day they delivered twelve large boxes to Emil's office, enough that he couldn't get in the door. Panicking, he hid them next door in the large-scale room. He didn't want the bosses to find out his big mistake. He's been taking them out, a few at a time, and giving them to various people and groups around the plant. Upper management is just as guilty as Emil, but we can't seem to convince old Dis over Der of it. He's still got forty-eight more to go. Did you want another one?"

Day with Dad

I began noticing an unusual situation in our house regarding the aluminum smelter. Because of my seniority in the plant, Joe, my elder son, had worked two summers for Alcan as a pot room repairman's assistant. At the dinner table in our house, occasionally Joe or I would mention some technical aspect at the plant, and we would get strange looks from Sharon or Skye.

Sharon had applied for a similar job and was turned down because she was female. At age seventeen, she was already developing into a strong feminist, and she was not enamoured of the Alcan policy that denied women high-paying jobs at the plant.

Driving the bus back and forth to Kitimat from Terrace, I had lots of time to think, and I realized that the three of them were cut out of this aspect of Joe's and my life. I decided to do something about it.

I got hold of Dell, in personnel, and told him I wanted to bring my nine-year-old son to work with me for one complete day. He'd come in on the bus with thirty-four men, change clothes in the locker room, go to my office and follow me around on my regular work.

At first it threw him, but then I could see a spark in his eye, and he said, "I'll have to see the big boss and work out the legal details and wrinkles. Regardless, you will have to sign a release form."

I said, "That's fine with me," and left it with him.

There are many hidden hazards in every direction at an aluminum smelter, so it's no wonder that it had to be thoroughly looked at. A week later Dell dropped into my office and said, "It's a done deal, but only as a trial. If it works, we may set up one day a year for fathers to bring their older children to work for a day."

I signed a hastily drawn up release form and told Dell I would bring Skye in the next Monday for the entire day.

I got up at my usual 5:30 a.m. so I'd be sure to have the bus at the pickup spot by 6:45. I woke a very sleepy kid and poured some breakfast in him. We walked to the bus, both with a lunch pail in hand, and I told Skye to sit two seats back behind me. I explained that each guy had his own favourite seat, and heaven help anybody who tried to take it.

Half the crew looked strangely at Skye as they got on the bus, but no one said anything. At the plant I let everybody off, then parked and locked the bus, and presented a copy of the release form to the security guard at the gatehouse.

My office was right in the potlines, so I had to get Skye a hardhat and goggles. He thought this was great. He had on a pair of boots—not steel-toed, but that was okay because I had brought salesmen in with normal shoes. I checked him for a watch, because some watches were ruined if you got too near the pot room due to the effect of the strong electromagnetic field. (It is also strong enough to ruin credit cards.)

We spent the day doing my normal work, plus a little extra that tourists don't experience. Everybody was overly helpful, glad to be part of this unusual situation. Toward the end of the day, one of the plant newspaper reporters, who had somehow heard about Skye's visit, dropped into my office. He wanted to do a private interview with Skye, and I turned them both loose in an adjacent office. The reporter seemed happy with what he got from Skye.

When it was time to go home, I didn't put him through the usual shower routine, with so many men there. As Skye boarded the bus for the trip back to Terrace, all the guys wanted to know his reaction to the plant. He was laughing and joking with them, and I think he felt he was one of the guys.

After we got home, he couldn't stop telling everybody what he had seen and done. A week later the Ingot, the plant newspaper, came out with the story, and was I ever embarrassed. The reporter had asked Skye his impression of what his dad did for a living, and Skye's answer, in large print, was "He goes from office to office, drinking coffee."

Needless to say, I noticed a subtle change at home. If something to do with the plant came up, I would try to explain, and Skye would interrupt me and do the explaining. I had no idea that he had absorbed that much information.

For the next few years they had a "Day with Dad," but it was gradually dropped. I think the company was the loser.

By the way, a few years later, Skye worked summers at Alcan and became a fully qualified "Pot Man." They also eventually started letting women work in the potlines if they could pass the strenuous tests. One woman I recall, due to her strong lower body strength, could do one of the toughest jobs in the potlines: changing over worn-out, 200-pound gas skirts on a hot operating pot. This was something most men could not do. She did this job for many years.

Wildcat Strike

In July 1976 the bad feelings of some of the more radical union members at the plant had exploded and we had a wildcat strike. Most of the union members walked out and set up a picket line on the road at the railway crossing a mile or so from the plant. The 350 members of management had to take over doing the work of 1,800 experienced men, operating the plant until the strike was over and the men came back to work. If we didn't maintain the pots, a tremendous amount of damage could be done to them. Basically, we were locked inside the plant.

Men who had never poked a pot on an anode effect found out what hot, dirty, sweaty work was all about. (See "Who's Stealing My Lumber?" in Chapter 4 for an explanation of the anode effect.) A nice young engineer, overdid it, working on the pots too long. He died from heat exhaustion.

I worked on the pots for the first three days, six hours on and six hours off, throughout the day and night. Then I ended up on Security and saw some very stupid, irrational damage that was hard to believe. I was stationed way up on the mountainside at the back of the plant, on top of a giant water tank that pumped water up from Anderson Creek. This water was critical to the electrical rectifiers; without it, millions of dollars of damage could be done to the rectifiers. Security had to take every precaution because we didn't know how crazy some of the young radicals might get.

I was given a set of binoculars and a radiophone and told to let Security know immediately if anybody went up to the dam and started doing anything. Away below me, to my left, was the Anderson Creek dam, with the pump and water intake up to the tank that I was sitting on. Below me and slightly to the left was the main plant road and Anderson Creek bridge. By the bridge, the strikers had a large campfire. Fifteen men were just sitting around it.

As I was sitting there, surveying everything with my binoculars, a worker I knew got up from the fire. Casually he picked up an axe and strolled over to a nearby telephone pole. He stuck the axe in his belt and slowly climbed the pole, then tied a rope around his waist and the pole. He pulled out the axe, cut all the phone cables to the plant, put the axe back in his belt, undid the rope and slowly climbed back down. He returned to the campfire and sat down as if he had just cut some firewood.

I immediately radioed in the damage and the fellow's name. Security took the information, but nobody came out to arrest anybody or repair the phone lines, and I began to wonder why I was there. (When the strike was over, that same fellow went back to work on the potlines as if he had done nothing wrong.)

After a few more days, things quieted down. The government sent in over fifty RCMP officers, who took up a position opposite the picket line. And the Alcan bosses decided that rather than trying to run the plant with 350 staff members, when it was usually done with over 1,800 workers, they would lower the power and run the pots on a sleep mode. They had already lost one man from heat exhaustion and overwork; they didn't want to lose anyone else.

One day, Ole, Harold and I were getting "bunkhouse happy" and decided to go out for lunch. We had heard the strikers were now letting cars through the picket lines, probably due to the presence of all those RCMP officers, so we jumped into my car—the one I kept at the plant all the time for emergencies or in case I had to work overtime, seeing as I was living in Terrace and would have no other way to get home if I missed the bus. Anyway, we got to the picket line and were stopped by one of our own millwrights.

A new man with the company, one of the real rebels, came toward my car with a smirk on his face and a gleam in his eye. He looked at the three of us in the car; then he climbed on my hood and proceeded to jump up and down, completely wrecking it. The RCMP just stood there, 100 feet away, and refused to be provoked.

I love all cars, but I had to sit there, clenching my fists, as this idiot wrecked my hood. The company had instructed us "Do nothing to cause a situation that the union can use to exacerbate the situation." It took a lot on my part to not open that door and kick the living shit out of the troublemaker. Needless to say I did not enjoy my lunch. After the strike the company paid to have the hood replaced. They also, when the men went back to work, transferred the troublemakers to other departments so "loyalists" couldn't extract any revenge.

A few days later, with the strike still on, I was back in my office in the dry scrubber building, looking after all the dry scrubbers, but I had a problem. Before the strike, Mary and I had planned a large wedding anniversary party at Lakelse Lake, 30 miles away, and we had invited a large group of people. I went to Andy, the maintenance superintendent and my big boss, and told him my name would be mud if I didn't get to my own party. He made arrangements, and I was flown over the picket line in a helicopter to the party and back, impressing the hell out of our friends. It's the little things that make a great company.

This was a wildcat strike, so there were some members of the union who felt it was wrong and who had volunteered to stay in the plant and work. One morning Ole and I were sitting in his office, having a coffee break, when one of his men came into the office with a really ugly problem. Because he had chosen to go against the union and stay in to work in the plant, he, like over 200 other workers, was considered a scab.

Howie was a big husky man who took no crap from anybody. He had gotten a radio call from his wife, who told him that a group of strikers were protesting outside their house. To top it off, as she was looking through the large picture window in the living room, Howie's best hunting and fishing friend threw a Molotov cocktail through the window. His wife put out the fire, but she was now nearly a basket case. She wanted to report the incident to the RCMP, but Howie told her not to. He would take care of it. Now he was asking Ole for permission to go home.

Ole agreed, but then asked him how he was going to get home. The union had closed the picket line, so we couldn't get out by car anymore.

Howie replied, "If you can't fly me out, I will go through the bush." He was a very angry, serious man.

Ole managed to get him flown out by helicopter. He came back the next day and reported in. Ole guardedly asked him if everything was straightened out at home.

Howie said, "She won't have any more trouble from that direction."

I was looking at Howie's large, calloused hands as he spoke. There were black, red and blue patches on his knuckles, with skin torn off. Somebody had taken a real working over that they wouldn't forget for a long time. I'm sure if Howie had reported the house damages, the company would have reimbursed him for them.

The strike ended after eighteen days. Nothing was really accomplished, just a lot of damage and stupidity. One of the stupidest events, in my mind, was when one of the garage apprentices, who worked through the strike, had somebody pour battery acid in his hat. Large patches of his hair never grew back.

The day after the strike was over, management sent a letter to all those who stayed in to keep the plant running. They were to receive $1,000 for every day they stayed. You can't just pull a switch and shut down an aluminum smelter, or, for that matter, pull a switch and start it back up. It would cost millions of dollars in damage to the pots. So the company could afford to compensate the people who'd worked through the strike.

A few days after the strike was over, everybody was walking out of the plant toward the main security gate as usual. The general foreman of one of the potlines, a feisty little guy, made a point of catching up to the president of the union and asked to shake his hand.

The president obliged and asked, "What was that for?"

The potline foreman said, "You see that nice red Ford F-250 truck sitting over there? Well, you paid for it, and I'd like to thank you for that."

Wild Goose

One beautiful late spring day, Skye and I decided to go fishing down Douglas Channel. There was an early run of spring salmon in Emsley

Cove. This is a large cove with one point jutting out into the channel, perfect sandy beaches and a stream running down to the saltchuck. The large cedar trees on the point were four to six feet in diameter and over a hundred feet tall, just like the ones in Stanley Park in Vancouver. They should have been preserved, but a few years later all those beautiful cedar trees were slaughtered. There went a nice park.

We were going around in large circles, trolling deep for salmon, when I noticed a large ship coming up from the south. Watching it, I noticed something strange. Even though it was probably four miles away, it wasn't following the recognized centre channel, like all deep-sea vessels, but was coming straight at us, very slowly. As we circled, the bigger ship seemed to turn slightly to keep heading straight for us. The peninsula and beach were behind us, and if it kept going on the same heading, it would land up on the beach. Ten minutes later it was a quarter of a mile away, still heading straight for us, though I also noticed that it seemed to be looping into every seaside cove or bay.

Surely that huge ship couldn't be fishing, I thought. But it was going very slowly for a ship of that size.

As it got closer I recognized it as a wartime corvette that had been converted for civilian use. Somebody must have money if they could afford to do that. It was over 250 feet long, probably with a crew of eight to twelve people.

We were still moving in a large clockwise circle. Maybe it was just my imagination, but as the ship got closer, it still seemed to be aiming straight at us.

When it got too close, I stopped my 18-foot boat and waited to see what happened.

The other ship slowed down and smartly turned to my left to come alongside little me. I could see the name "Wild Goose" was lettered on the side.

As it slowly passed, we heard all kinds of bells and the ship stopped. By then I was looking at the stern from 60 feet away.

I saw a sight I will never forget. Fully 25 feet above me, a man was sitting in a very big chair with a fishing rod in his hands. He had hooked a fish right by our boat.

We quickly got out of the way and watched this unusual spectacle. He played out a small blue-backed spring salmon, probably around ten pounds. A man dressed like a butler in the movies took the rod from the man in the chair. Then a sailor came and lowered the gangway, and the butler, or whatever he was, went down the gangway with a gaff. He reeled the fish in and gently put the gaff in its gills, while the man perched above us slowly got out of the chair and went over to the side rail. The butler lifted the fish up, and the man said quite clearly, "Yes, Tom, I'll have him for supper."

He leaned over the railing and yelled down to us, far below, in a very quavering voice, "How's fishing?"

I answered, "Not very good today, but better tomorrow."

He replied, "That's the spirit. Never give up."

At this point I had the strangest feeling that I knew this emaciated old man. His features and the sound of his voice were familiar. Whoever he was, he had to be wealthy in order to fish like that. He also had to be famous for me, in remote northern BC, to nearly recognize him. The ship then headed up the channel toward Kitimat.

Skye and I looped around a couple more times without any luck, and then I hooked something awfully strong. The water was clear and it was pulling like a dead weight. What had I hooked with our very slow trolling line? I pulled and pulled—I had never felt anything like it. When it was 15 feet under the boat, the fish turned over. All we could see through the seawater was what looked like a full-size white bedsheet.

I got to a point where I couldn't pull it up any farther or I would break the rod, so we rowed the boat close to shore, and I got out and pulled it onto the sandy beach. It was an eighty-pound halibut. What a thrill! I'd never caught a halibut before. The hook must've been on the bottom, where the halibut was hanging out.

We headed for home, and over supper I told the family about our unusual encounter on the water.

A few days later I was sitting in the kitchen after supper, reading the local paper. An article from Prince Rupert mentioned that the famous cowboy actor John Wayne had dropped by in his yacht, Wild Goose. Without knowing it, I had been talking to this popular movie star. (A year later I heard that he had died of stomach cancer. I guess that's why he looked so thin and frail when I saw him.)

CHAPTER NINE

The Invention of the Dry Scrubber Process

A Short Course in Aluminum Production   *185*
The Big Breakthrough   *185*
The Dry Scrubber Development Group   *187*
Campbell's Law   *193*
Dry Scrubber Adjustable Cyclone Vanes   *195*
Thinking Inside and Outside the Box   *200*
  Gas skirt burners   *200*
  Steam reaction   *204*
Suggestion Plan   *205*
A Diamond in the Rough   *206*

Dry Scrubber Basic Principle

To expose suspended alumina particles in a re-action duct, allowing them to adsorb and absorb gaseous and particulate fluoride and re-use the fluoride.

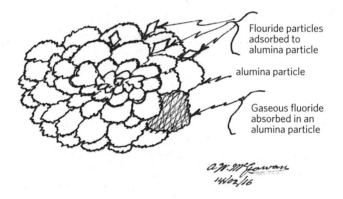

Ever since my participation in a two-week course on freethinking in groups, where "thinking outside the box" was put into practice, I have been a born-again fanatic in this area. And, during my time at Alcan, nothing was more exciting than the development of the 'out of the box' fluoride-collecting dry scrubber process, conceived by the near-genius inventor, Merlin.

A Short Course in Aluminum Production

Take alumina, mix it with the flux (fluoride plus cryolite), pass some very strong electrical currents through the mixture and the result will be molten aluminum, along with fluoride gas and particulate. End of lesson.

The Big Breakthrough

Fluoride is used in great quantities in the pots as a flux that helps convert alumina into aluminum. After conversion, the fluoride escapes as a gas that flows into the exhaust duct and away to the wet scrubber. There the fluoride combines with water, making a 5-percent solution of hydrofluoric acid. All this liquid goes to a settling pond for dispersal and is not recovered to be reused in the smelting process. This loss of fluoride costs many millions of dollars per year, not including the environmental damage (see "Corrosive Environment" in Chapter 2).

Merlin, an older, brilliant but very testy chemical engineer was down in the basement below potlines 3 to 5 one day when one of the pots had spilled a great amount of bath (that's the mix of fluoride, alumina and cryolite) into the basement. He picked up a large, brittle chunk that had unusual striations and colours throughout and took it back to his office among the potlines, where he placed it on his windowsill.

The window was slightly open, and after a few days Merlin noticed that this hard chunk of bath had become soggy and soft. Being of a curious nature, and having a smart, quizzical, questioning mind, he wanted to know how and why it had changed.

He went back down to the same location and picked another brittle chunk as a control sample. He took the two pieces of bath to the lab so they could do a comparison sampling and tell him the chemical difference between them, if there was one.

When the results came back, they showed that the soggy chunk contained a great deal more fluoride than the brittle one. Merlin mused about this. He knew that the air around the smelter was loaded with fluoride, which must have come through the open window. His conclusion was that the hunk of bath somehow attracted the fluoride, drawing it out of the air.

The big burning question now was how or why was this excess fluoride attracted to the chunk of bath?

Somebody at the lab finally figured it out. Basically, fluoride gas was being absorbed into the alumina particles, and the fluoride particles were also adhering to the alumina particles (see sketch above).

Thinking outside the box, Merlin realized this information could produce a huge technical and financial breakthrough if only he could develop a method to collect all the lost fluoride as it evaporated off the pots and flowed into the exhaust duct. We're talking about saving millions of dollars by reclaiming, and reusing, fluoride in every pot room around the world if Merlin or Alcan could develop a patented process. Plus worldwide patents. Plus big environmental improvements.

But as an aside, why did it take seventy long years handling millions of tons of alumina and fluoride to discover two of alumina's most valuable assets?

- It attracts particulate fluoride, which clings to the alumina particles.
- It absorbs gaseous fluoride.

Two days of experimentation in a laboratory would have revealed this information and saved the company billions of dollars by reclaiming fluoride. Not to mention reducing fluoride pollution by a tremendous amount.

The Dry Scrubber Development Group

A couple of years later, I was working as the pot room maintenance planner when a large pile of rusty, broken-down equipment—ducting, rotary valves, electrostatic precipitators, bins, air slides, etc.—landed on our doorstep from Arvida, Quebec, the site of another of Alcan's aluminum smelters. There were no phone calls warning us that this equipment was coming, no letters, no nothing.

Ole, the pot room maintenance foreman, made some inquiries through Bill, one of our mechanical engineers, and found out that a group had developed the concept for a miniature fluoride-collecting dry scrubber based on Merlin's concept. They had it built and installed it in the Arvida smelter, but they just couldn't get it to work. Some bright soul thought maybe we in Kitimat could get it working, so they shipped it out to us and literally dumped it in our lap.

A task force was now formed, the Dry Scrubber Development Group, made up of Bill as head, Gary as the chemical engineer, Ole and some of the maintenance crews to do the assembly, and me to help out where I could, along with my usual pot room maintenance work.

Bill was a quiet, mild-mannered, middle-aged engineer who knew how to listen, consider and act to get things done. Gary was a seasoned chemical engineer, heavyset, with an easygoing, flexible way of analyzing a situation. He also knew how to listen patiently and think objectively.

Ole was a heavyset Dane with a really flexible, quick-thinking, near-genius way of handling everything mechanical. Ole also had a secret life, which he finally let me know about a few years later. With his brain going at top speed all the time, he was constantly inventing things. One turned out to be a moneymaker, and he was quietly making and selling this invention on the side, shoving big money in the bank, for twenty-five years.

Bill and Gary got the whole crew together and explained the concept of the fluoride-collecting dry scrubber in full detail. Apparently we were to work with a concept design, but the whole crew was enthused when Bill and Gary stated that everything was experimental and they were wide open for suggestions. This was exciting, mind-blowing and challenging. The ideas came from everywhere and everybody.

In our spare time the pot room maintenance department assembled the material from Arvida and played around with it, modifying nearly every component, tearing out some of the pieces. Ole, the mechanical genius, was in his element, thriving on the stimulating challenge. I'm sure he had never had so much fun at work, and we had a hard time controlling him. His millwrights and welders caught the bug and were continually coming to him, or to me, with suggestions to solve problems or improve operations. If there was a day when nobody came up with an idea of how to get this miniature scrubber working, Ole was disappointed.

I don't know who designed our small two-pot dry scrubber model, but it wasn't long before Ole and his inspired crew, with the help of Bill and Gary, had it up and running on the two test pots. The collection of fluoride was a different matter, but with jiggery-pokery, lots of small changes, a refusal to give up, and the total elimination of the electrostatic precipitator (replacing it with a dust collector), they actually got it to collect some fluoride on the alumina particles.

The model was sent back to the Montreal engineering department so the people there could develop a large, fully functioning dry scrubber. It was installed in the line 3 courtyard to take all the exhaust gases from two potline buildings, 3a and 3b. The 200-foot-long, 5-foot-diameter contact duct handled 35,000 cubic feet of gas at 300 feet per minute, working at a temperature of 135°C. We got the first dry scrubber operating reasonably well, but it took all the brains of everybody concerned and many hours of brainstorming and modifying key components to get it up to 90-percent efficiency.

This full-size dry scrubber cost a little over half a million dollars to build, in 1973 money. Alcan calculated that within one year of operation it would pay for itself by collecting fluoride to be used again in aluminum production. Not to mention drastically improving both the air and water around the plant. These factors initiated a real push to

Our new dry scrubber office. Al McGowan at left and Ole Boye at right.

perfect the collection process so five more dry scrubbers could be built, using the modified dry scrubber on line 3 as an example.

By 1975 the Dry Scrubber Development Group had grown into a department separate from pot room maintenance, with our own office and shop. I was promoted to the position of dry scrubber technician, which inspired me to go back to night school and take a few courses in mechanical engineering. We were allowed tremendous leeway in an effort to perfect the process. We would modify and improve as we went along; then the draftsmen would be called in to make the necessary changes on the drawings. This was the opposite of normal procedure.

In 1976 we had a quiet celebration: Alcan now held a worldwide patent on the dry scrubber process. Direct credit was given to Bill and Gary. They were presented with acknowledgment plaques with their names on them, plus a silver dollar glued on as a token payment. I believe Ole should have been included in this group; because I was working with him intimately, I could see that he was the one who actually got it going, in part because of his refusal to give up in spite of the many challenges.

At this point, Ole got another one of his big brainwaves, which he bounced off all of us at one of our regular inspiration meetings. He and Gary had been playing around with the full-size dry scrubber. They were trying to develop a secondary recirculation system with a system of reused alumina.

The dry scrubbers worked because fluoride particles floating through the contact ducts for three or four seconds would attach to alumina particles. The alumina particles also absorbed some of the gaseous fluoride. Ole's question was, if one pass through the contact duct would collect some of the gaseous and particulate fluoride, could we collect more fluoride by reusing the alumina? They had conclusively found that alumina did continue to collect more fluoride, thereby increasing the collection efficiency. So Ole's idea was to figure out a way to recirculate the alumina that had already been used once and collect more fluoride.

We had all agreed that this concept should be tried, and Bill was going to take care of gaining approval from Montreal engineering, which meant getting formal approval, a budget and drawings for a secondary recirculation system.

A few days later I noticed that something weird was going on, though I couldn't find out what it was. More workers than normal were coming into my dry scrubber technician office and asking for requisitions, such as sections of air slides, rotary metering valves, pressure fans, electric motors and switches, etc. When I asked the tradesmen what account to charge these items to, they would say it was either to replace missing stock or for random repairs of various dry scrubbers. I had the feeling that I was being cut out of the loop, but with all the existing work I was doing, I'm afraid I didn't pay too much attention. The only thing I knew was that if Ole got an idea in his head to try something, he couldn't wait to get it done, regardless of the consequences.

Dry scrubber crew.

Bill dropped in one day and said, "You know that recycle process that Ole wanted to install? I've now got the money, $48,000, and the drawings. The engineer and draftsman are here from Montreal to explain how they see it being done. We're having a meeting in the works manager's office at two o'clock and you're invited."

That afternoon we all trooped into the works manager's conference room. The two "Wise Men from the East" were introduced and proceeded to tell us what we needed for the secondary contact process control system, describing the special parts that would have to be fabricated, and where and how to install them as per their drawings.

After the half-hour presentation, the meeting was open for

discussion about when and who and how to proceed. There was a short silence as everybody thought about this project.

Then Ole casually spoke up, and quietly said these earth-shaking words: "We have already built and installed the secondary contact process control system on the dry scrubber, and it is working very well."

There was a deathly silence as we all tried to get our minds around what had just been said. No drawings, no budget, no approval, but fabricated, installed and working successfully.

Ole sat there placidly, as if he had simply made a comment about the weather. We were all rather embarrassed. The concept of a "task force" had just gone out the window. And not one of us knew anything about it except Ole and the crew.

The manager finally broke the silence, saying, "Well, I guess this meeting is over. Nothing more needs to be said. I'd like to thank you two for your efforts, and you can take the next plane back to Montreal. I'd like to thank everybody for coming. Oh, Ole, could you stay behind? I'd like to have a talk with you in private."

I would have given anything to be a fly on the wall in that conference room for the next few minutes. By the next morning Ole was back to his cheerful self.

Bill had the Montreal drawings modified to conform to what had been installed, and we used them for the next five half-million-dollar dry scrubbers.

We were given carte blanche, and it was a great, stimulating, fun time for all concerned. I believe that one of the reasons we were able to make a success of it when Arvida couldn't was that they didn't have a complete maintenance crew like we did, with electricians, welders and millwrights at our fingertips. If one of us had an idea, we could go out that day and try it. They had to run everything through disinterested contractors. I also believe a large part of our success was the result of management giving us a loose rein. We accomplished a great deal, and normally we had a great respect for each individual's input.

We ended up putting in eight dry scrubbers, and with Ole's modification to accommodate repeated recycling we could improve collection of the gaseous and particulate fluoride from 90 percent with the dry scrubbers alone to 96 percent or more with the recirculation system, for a very substantial improvement in emissions and cost reduction.

Campbell's Law

Just as we were getting the first dry scrubber operating, a new member was assigned to our Dry Scrubber Development Group. We were now five, Bill M. was the head, Gary was the chemical engineer, Ole was the foreman, I was the dry scrubber technician and Bill Campbell was our new joe-boy, who helped everybody and recorded everything.

Bill came to us from the potlines, where he had worked since the plant started up in 1953. He went through all of the tough times as a pot room foreman, and "the powers that be" felt that after twenty-six years in the dirt and fumes, he deserved a break. During his years on the potline, he had made many good suggestions for improving pot performance. Now his health was slowly deteriorating, so we got him. In his new job as a technician, we could take advantage of his creativity. Lord knows, with the ongoing development of the dry scrubbers, we could use all the help we could get.

Bill had been in the army during the Second World War. He had been shot through the neck by a sniper and had half his jaw blown off, and he had been exposed many times to nearby explosions, so his body was riddled with small pieces of shrapnel. Over the years, some of these worked themselves out to the surface of his skin, so occasionally at coffee break he would pull one out by himself. Others created problems with muscles and organs, and he would have them removed at the Shaughnessy veterans' hospital in Vancouver.

Bill matter-of-factly explained all this to us, including what happened to his jaw. Apparently it was near the end of the war. His platoon was crossing a flat open field in Germany when he was shot by a sniper. He was immediately taken away to be looked after. The sergeant in charge figured out that there was only one place the shot could have come from: a large oak tree that they had passed by.

The soldiers all went to ground and slowly worked their way back to the tree, surrounding it. Looking up, they finally spotted a boy soldier, a member of the Hitler Youth, hiding behind a large limb. After four years of fighting, most of our soldiers had learned some basic phrases in German, and they called out to the boy to come down and surrender. No matter what they said, he refused, so they had to shoot him.

Bill's fighting days were over, and he spent six months undergoing reconstructive surgery in an English hospital.

In the Dry Scrubber Development Group, Bill was in charge of documentation, which meant that as we made modifications to the dry scrubbers, he would patiently rewrite the documents over and over again. Bill being Bill, he never complained about our harum-scarum habit of making myriad minor changes as we developed the dry scrubbers.

One day I came in to work and found Ole, the person guilty of most of the changes, sitting in his office with a downcast look on his face. I asked him, "What's gone wrong now?"

He pointed at Bill's office and said one word, "Look."

On Bill's wall over his desk, in six-inch-high letters, was a sign for all to see: CAMPBELL'S LAW ... IF IT WORKS, DON'T FIX IT.

With all our uninhibited enthusiasm for developing and improving the dry scrubber system, we had stopped considering the work we were loading on poor Bill. He got his message across, and for a while we were good little boys.

* * *

About two years later I was working away in my office when Bill Campbell came in and flopped down in a chair.

"I went up to the personnel office to see how much my pension would be if I retired," he said. "They worked it out and gave me this piece of paper." He handed it to me. I was shocked to see he was only going to get $450 per month at age fifty-seven.

I asked, "Bill, how come it's so low? You have put in more than nineteen years here."

"It's my own stupid fault," he said. "For some reason I didn't join the pension plan for eight years. Well, now I'm between a rock and a hard place. I know my health is failing, and I'd like to move back to Vancouver where we have a large, supportive family. But in my condition I won't be able to find a job there that I can do."

Very dejected, he got up and went back into his office.

I sat there a while, stunned, but then I had an idea. So I went over to Bill and asked, "Do you mind if I talk to someone about your situation?"

"Go ahead," he said, though without much hope. "It can't hurt."

When Ole came in, I buttonholed him. As soon as I explained my concern, he said, "Talk to our boss." He said that Bill had been with Alcan a long time and had all kinds of contacts within the company.

Bill M. came in the next day with his usual friendly smile. I cornered him and told him what I knew about Bill C.'s situation.

He looked at me and said, "I've known and worked with Bill all the time he has been with Alcan. That's actually why he is here, working with us. I'll see what I can do."

Two weeks later, Bill C. waltzed into my office with a big lopsided smile on his mashed-up face and said, "Take a look at this."

Alcan, via the Works Manager, had doubled Bill's pension.

Try that with any other company in today's world. In the 1970s people used to wonder why we all called Alcan "Uncle Al." I'd sure like to see some company today do something like this.

We gave Bill a grand send-off. A few years later I was in Vancouver on company business and looked him up. He was the manager of a twelve-suite apartment building, owned by his brother-in-law, with a small salary and a free apartment.

Dry Scrubber Adjustable Cyclone Vanes

In the mid-1970s we had two dry scrubbers operating beautifully, and the third one was being built. We were still doing a lot of fine tuning as the process went through its teething pains, and one day at our Joint Task Force biweekly meeting, Ole mentioned that they weren't getting enough coarse alumina particles back over to the secondary contact process system. Was there any way to lower the alumina collection efficiency of the cyclones?

Under normal circumstances, this statement would have been heresy. Cyclone manufacturers sell their product by touting how efficient the particle collection is. But because of our secondary process system, we needed the cyclones to be less efficient and to throw more alumina over to the dust collectors.

We tried playing with the cyclone's vacuum and feed, to no avail. Unfortunately, we couldn't go to any cyclone manufacturer because they would think we were out of our minds. They spend most of their time and effort making cyclones more efficient, and here we were trying to reduce the efficiency.

Bill, the head of the Dry Scrubber Development Group, came up with a world-renowned consultant who specialized in airflow and flew him in from New York at a cost of $2,000 per day. That was a lot of money at the time, but we were dealing with the possibility of recovering a lot more money.

When I started working on the dry scrubbers, I studied abrasive airflow and became fascinated by the subject, reading all of the little bit of information available. At one point I dreamed of winning the lottery and spending one or two million dollars on my very own wind tunnel so I could improve airflow on cars, trucks, motorhomes, etc. So when I heard this expert was coming to the plant, I wanted very much to meet and talk to him.

Finally the day arrived. Bill came into the office and announced that the great airflow expert was here from New York. Gary and I took him out to see our cyclone problem. The expert was a well-spoken, studious man. After we told him what we were trying to accomplish—basically getting a thicker coating on the dust collector bags to increase the fluoride collection efficiency—he seemed to understand our problem. He said he would send us drawings, so we could reduce the cyclone efficiency with a series of adjustable air spoilers and valves installed in the bottom of each of the cyclones to let in air.

A couple of weeks later we received the drawings. I ordered the four units to be fabricated by the Forge Shop, and at the next shutdown these giant valves were installed.

With great anticipation we restarted the dry scrubber. After adjusting and readjusting the spoiling valves, nothing happened. Back to square one, with $30,000 down the drain.

All of us, including the mechanical genius Ole, were trying to solve the problem. This got me wondering: If sophisticated, educated thinking couldn't solve it, maybe it was time to think outside the box with a simple, direct, "Rube Goldberg" method that might just work.

I got out my engineering book, with images of various cyclones of different shapes, and my drafting tools and went to work. It took me two weeks, but I finally got an idea and drew a sketch concept of adjustable deflector vanes, inserted inside the cyclone, that would upset both the centrifugal and centripetal effects. This would send larger alumina particles directly over to the dust collector. Because alumina is very hard (with a hardness of nine to diamond's ten), it will wear out any components it comes into contact with. To address this I designed the curve of the spoiler vanes so that a cushion of alumina would build up on them, instead of sliding across the surface and rapidly wearing them out.

I took my simple sketch of adjustable vanes to our next regular weekly meeting, and did they ever pooh-pooh me. Someone said they would wear out in a week.

As I mentioned, by now everybody had got into the act and was trying to invent something to reduce the cyclone efficiency. Bill C., our clerk, came up with the idea of a ring on the inside of the cyclone's lower cone. It was fabricated and installed, but didn't work. Somebody else suggested extending the outlet duct lower in the cyclone. Again I ordered a circular extension sleeve from the shop and installed it, but it didn't work either.

We spent many months and lots of money with this problem, but nothing we tried worked. Some experiments even had a detrimental effect on the operation. After each of these failures we would meet to discuss what to try next. I repeatedly brought up my adjustable vanes, and each time I was shot down.

Finally, desperate, Bill M. called a meeting with all concerned. It was a sad-looking group that assembled. They finally came off their high horses and agreed to try my idea, reluctantly appointing me to be in charge of installing my adjustable vanes in dry scrubber number 2.

The vanes turned out to work even better than I expected. We were able to control the cyclone's efficiency as we wanted, from 98 percent down to 84 percent. We retrofitted all the existing dry scrubbers with my design, and as each new one came online, they got my vanes as a standard. And due to the calculated curve of the vanes, they built up a heavy coat of alumina and didn't erode away in a week.

For some strange reason, I didn't receive verbal or written recognition for my invention. Maybe it was because I was only an engineering technician. Or perhaps I had somehow bruised some engineer's ego. I never did find out. Sometimes this engineer versus technician snobbery really pissed me off.

I submitted the entire package, recommending that Alcan patent it, but received a response saying that it wasn't worth the expense, even though the vanes were installed in dry scrubbers all over the world. Perhaps the drawing was just too crude.

* * *

Over the years I nearly forgot about the vanes. But in 1982 I was rudely awakened to the reality of ego, greed and ambition.

By this time I was working in the Reduction Development Department and happy as a clam. I had age and respect on my side, and a very satisfactory job trying to solve big problems for the potlines. I had two excellent bosses and sixteen intelligent fellow salaried personnel to work with.

One day I heard a voice from the past in the hall. Peeking out of my office, I saw Michael, a long string bean of a man, who was now on the international staff in Montreal. This is the dream job of nearly every Alcan engineer, as it sent him to any one of the sixty plants around the world to solve problems.

In talking to him, I discovered that he had just come from Sao Paulo, Brazil, where he was working on their dry scrubber installation project. In the midst of our confab, I casually asked him if they were installing my adjustable vanes in the dry scrubber cyclones. He gave me a very severe look, as if I were stepping on dangerous ground, and blurted out "no," then walked away. This bothered the heck out of me. Why was he so uptight? It wasn't like him. His attitude put a bee in my bonnet that would not go away. I wanted to find out what they did to those cyclones if they weren't installing my vanes.

In 1978, when I was the dry scrubber technician and planner, a Brazilian engineer named Juan came up and stayed with us for a month while he was studying the dry scrubber operation. Six months after I saw Michael, who should knock on my door but Juan. He was in Kitimat for some other business, and we went out for lunch together.

During the conversation he told me that they had installed dry scrubbers in the Sao Paulo smelter. So I asked him, "Did they install the adjustable vanes that I invented in the cyclones?"

He said, "Yes, and they sure work great. I didn't know that you had invented them. I didn't see your name on the drawings."

After what Michael had said, Juan's information surprised me. Why would Michael lie to me? I thought we were friends.

I asked Juan if he could send me the cyclone vane drawings when he got back to Sao Paulo. He said he would, and we bid each other goodbye.

Shortly after that the drawings arrived, and what do you know? Under "design by" was Michael's name.

None of this would have happened if Alcan had done what I suggested and registered a patent application. Another hang-up was that the company allowed only certified professional engineers to sign drawings. This "glass ceiling" between professional engineers and certified technicians prevented a lot of good ideas from being accepted.

I was hurt, and disappointed in mankind, but then I got mad and decided to do something to right a wrong. I gathered up all the eleven-year-old records I had that were evidence of my work on the cyclone vanes: meeting minutes, work orders and, most importantly, my dated sketches. Then I got Gary, a professional engineer and one of the original members of the Dry Scrubber Development Group, to write a letter confirming my invention and its local success.

I took all this data to my boss, who reviewed it and sent it, with positive recommendations, to our department boss. I was not privy to any further actions or recommendations, but on June 8, 1988, they presented me with a thank-you letter and a cheque for $2,000 as a bonus for the work I'd done.

Although, strangely, nothing was said about the cyclone vanes in this letter, nevertheless I really appreciated it. With the bonus money I took Mary out for a very fancy supper, and I bought myself a beautiful set of golf clubs with the remainder. Many times when I picked up those clubs I had pleasant memories of the three people who helped me receive recognition for something that I had invented. I don't know if it was ever patented, but it was used in a lot of dry scrubbers.

Thinking Inside and Outside the Box

Aluminum smelters have been around for over a century, and there are probably more than 100 smelter facilities around the world. A burning question I have often asked myself is "Why did it take so many years to discover that alumina absorbs gaseous fluoride and adsorbs particulate fluoride? With at least 200 chemical engineers involved worldwide, how come nobody discovered this fact earlier?" A couple of simple comparative tests would have revealed this and produced the breakthrough before the 1970s—though then I wouldn't have been able to work on it.

Here are a few more examples of "outside the box" thinking.

Gas skirt burners

Each pot had at least one gas skirt burner. These are supposed to burn off any of the combustible gases coming off the pot before they get into the exhaust ducting or into the dry scrubbers. If these gases weren't burned off, they cooled and plugged the burners with soot. When this happened, the pot man had to physically clean the burner and try to relight it, a very dirty, difficult job.

As a technician, I was given this problem to solve. I studied it, trying to figure out why the burners didn't always eliminate all the gases coming off the pot, and I talked to everybody involved, the superintendents, engineers, potline supervisors and especially the pot men who had to maintain the burners. I got a few good suggestions, but I needed more expert advice.

One of the many hard lessons that I had learned about solving problems was that I needed to go to the source, which in this case was the pot men who had to clean and maintain the burners. Another lesson was to pick out the most negative, disgruntled people to see if I could find the answer to the problem with a form of reverse psychology.

In this case I had a real dandy. Harry couldn't say anything good about anything, but he was a thinker, even though he didn't know it. I went out into the potlines and found Harry, then sat down with him, told him my problem and explained that I didn't know anything about burners. He was the expert.

Harry mumbled and grumbled and ended up taking me over to the first available pot. He got a scraper and showed me where the soot actually built up, muttering, "It's a pain in the rear to keep cleaning it out every bloody day."

In these exchanges with the workman, you have to ask stimulating, argumentative questions and then ask the killer question. In this case, "What would you do to the burners if you had the chance?" At this point, if you have stimulated him enough, grab your hat and hang on. It would even be nice to have a hidden microphone, because you're going to get the good, the bad and the ugly. But somewhere in there you usually find a brilliant answer to your problem.

Harry did give me a lot of good information, so I thanked him and went on my way. But I needed to get Harry's comments clarified by an expert. I checked around the plant and, strangely, discovered that we didn't have an expert in gas combustion.

This was a time when oil and gas prices were sky-high, and everybody local was looking at alternative ways to heat their homes, preferably with wood, which was abundant in our area.

Hans was a qualified combustion expert who lived in the town of Smithers, 150 miles away. He had stuck his neck way out and started a little factory, where he made wood stoves and heaters. I had been out to look at his factory, and especially his heaters, while on a hunting trip and had been very impressed. He had six men working for him and produced extremely efficient heaters that he sold throughout northern B.C. Of most interest to me was the rare-metal afterburner he had developed to burn off any of the remaining fumes.

I talked to my boss, and he gave me the okay to hire Hans as a consultant. When Hans came over from Smithers, I briefed him on what I had found out. I also got him to talk to some of the pot men about their various problems. Hans took our burner drawings back home and modified them to eliminate the dead spots in the burners, which were the main cause of the soot buildup.

I took the drawings down to the Penticton foundry in the Okanagan and got them to send us a few modified burners. These were installed, and they improved 50 percent of the burner problems. We made them the standard burners. Is this thinking outside the box or not?

Our plant paper, the Ingot, did a write-up on the problem and how I solved it, and somehow the president of Alcan saw this article and sent me a personal letter, congratulating me on "thinking outside the box" to solve a problem. It's pretty nice to get a personal letter from a man who has 60,000 people working for him, and I treasure that letter.

About six months later, out of the blue, I was called into my boss's

Visit to Kemano with distinguished company.

office. He bluntly asked me, "How would you like to go to Kemano in a helicopter?"

Without any hesitation I said yes. I had never been to our giant underground electricity-generating facility.

My boss then said, "You will be going in distinguished company, along with the company president."

We met on the landing field. He was a quiet, fit, middle-aged French-Canadian. Along the way, he quietly asked me the damnedest questions about Alcan, my job and my family, and we got along just fine.

I've often suspected that he was feeling me out for another job somewhere else. But I put a stop to that and told him I was so involved with my life here that, reluctantly, I couldn't afford to move. Anyway, it was nice to be recognized. You never know where recognition or praise is going to come from.

Steam reaction

During the formative years of the dry scrubber development, dry scrubber number 3 was our favourite guinea pig, probably because it was close by our courtyard office.

It was my job as dry scrubber technician to record the efficiency of all the dry scrubbers. I had a large graph pinned on my office wall and recorded all these results daily. I happened to look at the statistics one day in the fall rainy season, and I noticed that during the previous three weeks of steady rain, the fluoride collection efficiency was up from 90 to 96 percent.

Could there possibly be a relationship between humidity and alumina's ability to absorb more gaseous fluoride? On my own, I bought a barometer and a humidity gauge, and started recording the ups and downs of humidity as it related to fluoride collection efficiency.

After three months of recording, I was sure there was a relationship. I called in Gary, our chemical engineer, and drew his attention to this phenomenon. He showed only a mild interest, so I mentioned it to

Ole. He lit up like the Fourth of July, and we ended up doing a slightly dangerous experiment all on our own.

We installed a steam line at the head of the fluoride contact reaction duct and slowly turned on the steam. Result, collection efficiency instantly went up. There was one risk: too much steam could gum up the works of the dust collector bags.

We notified Gary of this important breakthrough, and we ran with the steam operating for the next few months.

Gary wrote it up, but when we didn't hear anything back we finally shut the steam off, much to my disappointment. I kept bugging everybody—Should we expand steam to all eight dry scrubbers? Why didn't we patent this idea?—but my idea went into upper orbit, and we couldn't get an answer.

Three years later a dry scrubber technician in Sweden noticed the same phenomenon. He quit the company and patented it worldwide. Now if we wanted to proceed with steam, we would have to buy the patent from this fellow. Did this make sense? Who dropped the ball and why? Being the original inventors of this process, I thought Ole and I would at least hear back with an explanation of why we didn't patent it, or with an apology. (Coincidently, a few months after the Swede patented the process, all Alcan personnel were asked to sign a secrecy agreement that said all inventions they came up with while working for Alcan were the property of Alcan. This applied to anything related to Alcan for five years after they left employment.)

This was one of many things that can be frustrating when working for a large industrial company. Ideas seem to disappear in an upper-level quagmire, never to be seen again.

Suggestion Plan

At one point Alcan had instituted a Suggestion Plan, which offered employees money for good suggestions. As someone who strongly believed in the concept of thinking outside the box, I thought this was one of the best morale boosters that had ever come down the pike, particularly for those employees who were inclined to have good, but often weird, ideas for improvement.

One day, Gene, our best brick mason, came slowly through my office door. I asked him what he wanted, and with much hesitation he told me of a brilliant but extremely dangerous idea he'd had. He'd come up with a plan for replacing the Zircon block and plate in a casting department furnace without having to shut the furnace down.

Here is the background to Gene's suggestion.

The giant casting department furnaces were three feet deep and held upward of 30,000 pounds of molten aluminum metal at 1,900 degrees Celsius. To move this molten metal out of the furnace to the next step of the production process, you had to have some means of controlling the volume of metal pouring out of the furnace. What we had was a large, expensive, custom-made Zircon Spigot Block, with a small hole in the centre of the block and a plug that fit the hole. This block was attached to the lower side of the furnace.

Over a period of time, the metal flowing out would enlarge the small hole in the block until it became too big for the plug. When this happened the whole operation came to a halt. The furnace had to be drained and cooled off. Then the outside steel plate holding the Zircon block was removed from the furnace, the worn block was removed from the plate, and a new block was reset in its place. After the steel plate was reinstalled, the furnace was restarted.

Unfortunately, to heat a furnace from square one, or to cool it down to a point this operation could be performed, took at least a couple of days. And furnace downtime was calculated to cost at least $1,000 per hour.

Gene's revolutionary idea would let us replace the Zircon block without shutting down the furnace and would take only six hours as opposed to forty-eight. This idea would produce tremendous savings—if it worked. Personally, I was horrified at the risky concept Gene had developed, but I was willing to go into outside-the-box mode. To my surprise, Casting was willing to give the idea a try. We waited until a

Zircon block was worn. For safety's sake, we partially drained the furnace before setting up the fans to freeze the metal. There was a large crowd gathered when Gene slowly and carefully removed the worn Zircon block. The frozen metal around the hole was still a dull red colour, and just a wee bit scary. If that frozen lump of metal sprang a leak, we would have an extremely costly mess on our hands, never mind the people getting serious burns.

Gene put in a new gasket and installed the new Zircon block and plate. We restarted the furnace after a downtime of possibly six hours, versus forty-eight.

This new procedure became the standard practice for replacing Zircon blocks in all the smelters. Gene got a small hunk of well-earned cash, and the company saved big bucks on downtime for all the furnaces throughout the company.

A Diamond in the Rough

When I first met Harold back in 1954, he was the assistant chief steam engineer for the smelter heating system. I was painter gang leader, and my team did a lot of work for him. All the new steam plants and large air compressors had to be painted, with all the piping color-coded.

Eighteen years later, Ole, and I were in charge of the dry scrubber system throughout the plant, and Harold and the new steam plant were our neighbors when we moved into our new central building. Harold ended up in the office beside me, and next thing we knew this very gruff, taciturn introvert was having coffee with us and going uptown with us once a week for lunch.

This is when we really got to know what made Harold tick, I don't think I have ever met a more complicated human being. He was a product of the many environments he had been in. He must have come from a good home in a small town in Ontario, because he could play the piano beautifully. For some reason that he would never disclose, he quit school in Grade 8, and at age fifteen he joined the Merchant Navy during the Second World War. He had two merchant ships sunk under him, one just outside Sydney Harbour, and one on the North Atlantic convoy run. It was this latter sinking where a broken steam line blew out his left eye.

Harold was one of those rare men who for some reason displayed unusual strength in a normal-sized body. This led him into fights with men twice his size, which he always won. An enterprising friend in the Merchant Navy talked him into becoming a prize fighter, with the two of them splitting the profits. Since they were sailing all over the world during and after the war, they were frequently in ports where they weren't known, so the prize-fighting pickings were easy. Again, Harold easily adapted to this way of life, with its fighting, drinking and wild wild women.

He also developed another questionable occupation. He got involved in the black market, buying restricted items in one port, hiding them on the freighter and selling them in another country.

He and his partner were in a small South American country, promoting a fight and betting on Harold, when things turned ugly. Harold had a bare knuckle match with the local hero, and when Harold punched or gouged the local hero's eye out, it did not go over very well with the local population or the judge. He got six months in jail under unusual conditions. Harold was shipped 60 miles inland, to a wide open jail, where escape back to the coast was nearly impossible.

Each morning inmates were given a small breakfast; then their job, if they wanted any lunch or supper, was to drain the milk from a certain number of coconuts, split them and take out the meat. To get more food they were turned loose in the town. If they couldn't find work to earn money for food, their options were to beg on the sidewalk or eat coconuts.

Because they were mostly eating coconuts and nearly starving, the inmates' stools became little pellets, like a rabbit's. This meant they didn't need to wipe their rear. Each man was given only one small square of toilet paper per day, which they saved up. When a new prisoner showed up, requiring the normal five to six squares of toilet paper

Al and friend Harold at Alcan retirement party.

per day, they would sell their stock to him to make some money, even though after a few days the new prisoner would no longer have to wipe his rear.

After serving his six months, with fellow inmates stealing everything from him, Harold had to hitchhike and walk back to the coast, where he contacted his old ship and asked the captain to send the money so he could get back on board.

About this time, Harold finally realized that he was not going anywhere with his itinerant way of life and decided to quit the sea.

Over a period of years as chief steam engineer for Alcan in Kitimat, he built an airplane from the ground up. He took it to the local airport and had a pilot friend take him up on a test flight. He got the plane certified as air-worthy and then sold it.

Stupid me, I asked him why he didn't fly it for a while and enjoy it after all the years of work he had put into it.

He just looked at me compassionately and said, "Alan, I don't like flying. I just like the challenge of building planes. Besides, they won't let me fly with only one eye."

Ultimately he built two planes, each of which took over 5,000 hours and seven years to complete. Harold got a great deal of satisfaction building them, and he felt the thrill of accomplishment when he saw them fly.

It took a long time for Harold to learn to trust us, but once he did he slowly opened up to us in some very strange ways. It was as if he had arrested development and had no idea how to live normally.

One time Ole and I were discussing diets with various foods. We mentioned something about a balanced diet, and Harold interrupted us. "What the hell are you guys talking about, a balanced diet?"

Ole replied, "A balanced diet keeps you healthy and strong."

Harold then asked, "What do you eat with a balanced diet?"

Ole replied, "Oh, you know, the usual things, potatoes, vegetables, meat and maybe cake for dessert."

Harold asked, "Is this supposed to make you healthier?"

And Ole said, "Yes, it should."

"This sounds real good," Harold said. "I'll give it a try."

Four weeks went by, and during that time the subject of a balanced diet didn't come up. One day Ole remembered our conversation, and during our coffee break he asked Harold, "How is the balanced diet coming along?"

Harold replied, "It's working great. It's a lot easier getting meals ready now."

Knowing Harold, Ole hesitantly asked him, "How come preparing a balanced diet is easier than the way you prepared your meals before?"

Harold looked at both of us with a straight face and said, "For week one I cooked up a mess of mashed potatoes. For week two I cooked and ate vegetables. For week three I cooked up a big roast beef, and for week four I bought cakes."

We nearly fell over laughing.

Harold retired more than ten years before Ole and I did, and every Wednesday we kept up the practice of meeting for lunch uptown. There was only one fly in the ointment. Every Wednesday at 10 a.m. it was my job, and I mean job, to phone Harold at his home and remind him that it was Wednesday and we would meet him at a certain café.

The reason for the phone call was, believe it or not, that Harold, all alone in his large house, would get so involved with building an airplane that he would forget what day of the week it was, not to mention what hour of the day or night it was. Sometimes when I phoned he told me he had just worked thirty-six hours straight without eating.

Usually the conversation would go like this:

Alan: "Good morning, Harold. It's Wednesday. We are going to meet you at twelve noon at Helen's Cafe. Is that okay?"

Harold: "Who's phoning me?"

Alan: "Harold, it's me, Alan, and we're going out for lunch today. Are you coming?"

Harold: "Why are you phoning me at ten at night? Besides, it's only Monday, for Christ's sake!"

Alan: "It is Wednesday, and we are at work, and it's ten o'clock in the morning. What have you been doing?"

Harold: "I'm working on the curved leading edge of the wings, and I guess I've lost all sense of time. Are you sure it's Wednesday? Because I think I started on this on Monday."

Alan: "Well, Harold, get cleaned up and meet us for lunch."

Harold: "Okay, but are you sure it's Wednesday? You wouldn't be trying to fool me would you?"

Alan: "No, Harold, we will meet you at 12 in the flesh."

I went through this frustrating process nearly every Wednesday for the next ten years.

Harold, being an old sailor, knew how to drink and how to hold his liquor. But one night he hit the Canadian national headlines for drinking. He was at the local Legion on a Saturday afternoon with a bunch of the guys, and he had many drinks of rum. Just as he was getting up to go home for supper, somebody ordered a double rum for him to chug-a-lug before he left.

Harold being Harold, he thought nothing of it and tossed it down. He walked out the door, got in his car and within a hundred yards was caught in a police roadblock. He registered 2.5 percent on the breathalyzer, which meant legally he should have been dead.

We never let him forget that he hit the national news with the headline "Man Driving, Legally Dead."

CHAPTER TEN

New Challenges

The Aluminum Handshake and
   a One-Man Strike   *215*
Slowdown Year   *217*
House Building   *220*
Invention of the Curl Stick   *222*
Reflections   *225*

My 30-year award.

Early in my working life I read something that hit home: after seven years at a particular job, workers become stale and should request a transfer so they can become reinvigorated again. In my thirty-six years with Alcan I followed this principle and was never disappointed: I spent seven years as gang leader for painters and glaziers, seven years as Building Trades Maintenance Planner, then ten years as a dry scrubber technician. As I entered my final years with the company, I had a dream job in the Reduction Development Office, taking on large potline projects that might carry on for years before they were completed. If I wanted to travel to see some other Alcan smelter or the competition to figure out a problem, I had no hassle getting permission to go. The other members of the department were an intelligent group, and we worked extremely well together but changes were coming.

The Aluminum Handshake and a One-Man Strike

We were heading for a minor recession, and the company wanted to trim some of the salaried employees. Upper management sent a letter to salaried staff, telling them that they wanted to reduce staff by up to eighty people and would give long-term employees a bridge pension, a generous retirement pension and improved health benefits. This offer was so generous that the plan backfired. Basically the managers hadn't taken the "Kitimat Syndrome" into consideration: due to bad weather and isolation, 98 percent of the population wanted to be somewhere else, particularly when it came time to retire (thank God this phenomenon has finally, after all these years, nearly gone away).

First off, well over half the staff wanted the package and applied. This shocked upper management in both Kitimat and Montreal, and they had no idea what to do.

Second, some foremen and general foreman had, over many years, built up unbelievable amounts of banked "call-in time." Some had up to two years in the bank. This meant the company would have to pay them another two years of salary, over and above the retirement bonus. Who wouldn't want to apply for this wonderful unplanned bonus?

The result was that the company had to pick and choose who would receive the retirement package, and they refused to give it to many of the long-time employees—in part because they were too valuable to the company and Alcan didn't want to lose them, and in part, I'm sure, because they didn't want to pay out for the banked time.

This meant there were a lot of valuable, long-time employees who were now also very unhappy.

Around this same time, Ron and I formed a social group, "The Pioneers of 54," made up of the men who had started at Kitimat when the plant opened and were still there. There were thirty-seven of us, and we held meetings and social events. I still have our large banner at home, and twelve of our custom-made hats. We had the company's complete list of birth and service dates, and we were well aware of who was eligible for the retirement package—based on the criteria of being over fifty-five with thirty years of service—but had not received it. In a way, the Pioneers of 54 became a mild protest group, made up of people who didn't get that "Aluminum Handshake."

A typical example of this was Joey, whose case evolved into a major morale problem. Joey had hired on in 1954 when the plant started up, same as me. He served time as a pot room worker in the tough early few years. Then he was promoted to foreman. He had done excellent work over the years in many different jobs, collecting many bonuses and letters of appreciation.

Joey was a tall, slim, happy-go-lucky man with a good, intelligent way of handling manpower. He had worked his way up to the position of general foreman, with 135 men under him. But now his wife was sick with cancer. She had to go and live in Vancouver for treatment, which put a huge strain on Joey, who wanted to be with her.

He reasoned that he fit the bill for the Aluminum Handshake, being four years over the eligibility level of fifty-five years old, so he made a tentative verbal inquiry, asking if there was any chance that he would be considered. Management informally told him that he was "too valuable" to be let go. I applied and was told the same thing, and I reluctantly accepted it.

However, Joey told me that he thought it over and decided to make a formal written submission to be considered due to his record and his wife's condition. He was formally refused, on the same grounds: he was too valuable an employee.

So Joey went on a one-man strike. He came to work as general foreman with 135 men and three foremen under him, but he refused to do anything in a high-priority department.

Because Joey refused to continue to work in a key production job, they transferred him to a nice cushy job that nearly ran itself. There, he proceeded to do less and less each day, sitting with his feet up on the desk, setting a bad example for everybody with his conversion from being highly productive to a dead standstill.

Negotiations with the personnel manager crawled along, and it wasn't till Joey had been on his one-man strike for nearly a year that they made him "an offer he couldn't refuse." That was all he would say, as he had signed a confidentiality agreement and was sworn to secrecy.

What a strange situation! After knowing him as another Pioneer of 54, I wished him well in retirement. We bid him adieu, and he went to rejoin his wife in the sunny south.

Slowdown Year

After the great upset over the Aluminum Handshake, everybody who was left settled back down into a nice working relationship.

One day I had something to do in the pot room area. I took the stairs down from my upstairs office in Reduction Development and decided to take a shortcut through the Forge Shop. I was halfway through all the racket when I heard a strong voice behind me, calling my name.

Len, the Forge Shop planner, had come out of his office and was waving some papers at me. I turned around and retraced my steps, and he signalled me to come into his office.

I followed him in, and he closed the door to keep out the continual racket. We both sat down and he thrust the papers he had in his hand at me, saying, "You realize you are now working for a dollar ninety-eight per hour?"

Taken aback I stammered, "What are you talking about? I'm making $22.50 per hour."

Len shot back, "You and I came here thirty-five years ago. Our careers have been nearly similar. If you retire next year at age sixty and take your full company pension, plus your old-age pension, plus your Canada pension—add them all up and divide them by the number of hours you're working each year and you're making $20.52 per hour, which means you are now working for roughly a dollar ninety-eight per hour. At least that's what I figured out for myself. Does it make sense to keep working? I for one am handing in my notice and heading back down to the Okanagan."

My head was in a whirl as I staggered out of his office, across the passageway and back up to the quiet of my office.

I knew Len had done a lot of groundwork to come up with this radical concept. It took me a week of checking with the personnel department and doing my own calculations to finally admit that Len was right. Our figures were nearly identical.

It was time to consult with my better half.

At this point we were dealing with empty nest syndrome. Joe, our eldest, was long gone; Sharon was living in Vancouver, doing her thing in radio and TV; and Skye, our youngest, was in Victoria, learning how to be a teacher.

I tentatively broached the subject of retirement with Mary, explaining the weird situation we were in. Without batting an eye she said, "Well, if you want to retire, and it makes money sense as you say, go ahead and retire." But a moment later she continued, "Knowing you, you're going to have a lot to do in retirement to keep you happy. I don't want you hanging around the house, bothering me and my friends."

Both of us had taken a course on retirement, and I knew where she

> SKIRTS COATED & CHASED, BURNERS BURNED, SCRUBBERS SCRUBBED, PLANNERS PLANNED, PAINTS BRUSHED, BUSES DRIVEN, LAND LORDED, COWS POKED, ROCKS PROSPECTED
>
> He's done it all, but now it's time to quit.
>
> ## Al McGowan is retiring
>
> Alcan Tour Building
> March 29, 1990.
> Starts at 4.00 p.m.
> Cost is $ 5.00 including refreshments.
> For more information contact
> Harry Eisenberger 3368
> Norma Hall 3715

was coming from. I needed to draw up a long-term schedule of what we, and I, would be doing in retirement. We decided I would retire in a year, which gave me lots of time to work out that slowdown schedule.

The next big step was going into my boss's office and giving him notice that I would be retiring in one year. He accepted it gracefully, and over the next year I slowly put all my long-term projects in order and phased various people in on all the particulars of whichever project they were going to take over. I found I was rather enjoying myself during this period, strangely enough, and I became more assertive about everything.

Harry was given the job of organizing my retirement party, and he did a great job. Heber, an old Haisla friend, who I'd taught how to paint, came to the event and presented me with a Robert Stewart wall carving. All in all, it was a nice way to go.

And what was on my slowdown schedule?

Number one. I decided to sell off my twenty-four rental units, as they were starting to become too much work.

Number two. I had to learn how to slow down, and I decided to do that by renovating a charming old house that my company, Skeena Estates, owned. It took me a year to reduce my hours of work per day from ten hours to two or three, but in the course of doing that I completely rebuilt the house: raised the sagging floor, installed a shake roof, replaced all the galvanized iron plumbing and installed a new modern electrical control panel. I had no sooner finished the project than I was offered a price 20 percent above the house's value. The offer came from a developer who was more in the know than I was. I sold it, and a week later it was torn down, replaced by two six-unit apartments—much to my chagrin, as I loved that old house.

Number three. To further learn how to relax, I bought a golf cart and joined a group of old geezers on the golf course once a week. I had always wanted to learn how to play golf, but I had always been too busy with everything else.

It seemed to work, as both Mary and I survived that first year and carried on into retirement.

House Building

Mary's mobility was slowly deteriorating, what with her bad arthritis and our house having three floors.

This gave me a perfect opportunity to build a house, something I'd always wanted to do. I had built sheds and shacks, I'd maintained all these rental units and I'd certainly spent a lot of time making repairs to somebody else's mistakes. But I'd never had a chance to build a complete house from the ground up.

So I got the plans for two different sizes of one-floor house that would fit on a lot I owned. I then made little icons of all our furniture—tables, chairs, beds, etc.—drawn to scale so they would fit on the floor plans. This would give Mary a better idea of which one of the two we should build.

Our new house plus workshop, 1993.

One plan was 2,000 square feet; the other was 2,500 square feet, both on one floor. I had both plans laid out on the dining room table. I showed Mary all the miniature furniture and the layouts.

Barely looking at the drawings, she asked, "Which one is the larger?"

I pointed to the 2,500-square-foot one. With no hesitation she pointed to that one too and said, "Build it."

By this time I had been retired for over two years and was gradually slowing down, so I developed a house-building philosophy: Get somebody else to do the rough major work. Then I'd do the rest myself to save money.

The major contractor had no sooner broken ground than I received the bad news that I had cancer in my right kidney. No doubt this was the result of my repairing and rebuilding seven wrecked cars, and commercially painting over ninety in small cramped garages, inhaling all the fumes. Down to Vancouver I went to have it cleanly removed.

By the time I came home, the contractor had poured the basement and was framing the house.

I tried to help, but being only two weeks out of the hospital I promptly broke the interior stitching and had to take treatment for it. Ever so slowly I recovered, though it took me fully a year and a half to regain my energy. Every week I did a little bit more work on the house through the pain. This worked well, but it still was nearly two years before I finished the house, doing the wiring, plumbing, painting and flooring.

We bought all new furniture and moved into our retirement home. Now Mary just had to push a button and the garage door would open so she could drive in, get out and walk with a cane straight into the house, with handrails.

After the house was finished, I felt it was my turn, so I built a large workshop in the backyard. That is where I have spent many thousands of happy hours "doing my thing."

Invention of the Curl Stick

Now that the house was finished I was golfing, fishing, riding with my motorcycle friends. It all occupied my time but I didn't feel challenged. This is always a danger time for me. When I have spare time my mind wants to keep busy, and my body needs more to do, and the inventing bug creeps up and grabs me by the ass.

One activity I took up in retirement was curling. By my second year in the sport I knew the whole seniors curling group. One of the members, who was on my four-person team, was Ray, a man older than I. He had lived and breathed curling for all of his life, and he helped me learn the finer points of the sport.

One day we were out on the ice, and it came to be Ray's turn to throw his rock. With some hesitation he kneeled down on the ice and delivered his rock, but then he couldn't get up again. Two of us had to go over and help him get back up on his feet. He was so embarrassed! He had been a very independent person all his life, so it was hard for him to admit he needed this help. He walked off the ice, not saying a word to anyone.

We carried on and covered for him, with nobody saying anything. I was the new one on the team, so I didn't want to bring it up. We finished our game and still nobody said a word about Ray walking off the ice.

The next week I arrived at the curling rink to find Ray sitting on the bench behind the glass. Nobody was talking to him. I thought this strange, so I went over and asked, "Aren't you going to curl today?"

He looked at me and said, "If I can't get down on the ice to curl, I'd rather watch."

Now, technically he could get down. He just couldn't get back up! But this situation bothered me so much, I went into "out of the box" thinking. I was determined to invent something to help him so that after forty years of curling he could continue to take part in the sport.

I went into my nice new shop and carved out a wooden curling stone with a regulation handle. I put a long piece of carpet on my shop floor that I could slide the stone on, and I set to work figuring out what to do. First I did a breakdown analysis of the problem.

One. Ray had to stand up and curl, so that meant some form of extension from his arm to the curling rock.

Two. Whatever it was, it had to be able to put curl on the rock.

Three. It had to smoothly release the rock.

Four. It had to be adjustable in length to fit various people.

Five. It had to be able to clean the bottom surface of the stone before delivery.

Bill Reynolds, my neighbour, and I brainstormed the concept. Over five days of trial and error we solved all five problems and developed what is now a patented invention called the Curl Stick, which is used around the world.

I couldn't wait for our regular Tuesday morning seniors curling session. When the day finally came, I very excitedly took the curl stick down to the curling rink.

There was Ray, gloomily sitting behind the glass. I went up to him with the curl stick in both hands and presented it to him as a gift, saying, "With this rig you now can curl standing up."

He took one look at it and said, "I wouldn't be seen dead curling with that thing. Now get it out of my face."

This made me mad. I told him, "All right. If you won't use it, I will." And I did. I used it throughout the whole game.

When it came my turn to throw my rock, I placed a small rubber-backed carpet on the ice. Then, with the end loop on the curl stick, I lifted the 32-pound rock onto the carpet by the handle, gave it a few twists to clean the bottom and proceeded to curl, to the intense interest of all the players but Ray.

Nobody had ever seen anything like it, and when the game was over, everybody wanted to try the curl stick, with great enthusiasm. In fact, I got orders for two more, and McGowan Curl Sticks was born.

I spent the next several years developing, patenting and marketing the curl sticks at rinks across the country. It was a lot of work but a lot of fun too. However, after they became accepted and used widely,

Testing first Curl Stick, 1992.

some other people started making their own (with slight modifications) in their garages and I couldn't fight them off with my patent rights. I eventually closed the project down and refurbished my remaining curl sticks into canes. These I donated to a charity that ran an orphanage for disabled children in the Dominican Republic. I accompanied the shipment to the orphanage and saw them distributed to the children myself, which was a very rewarding experience.

Reflections

When I look back objectively over the many years, I realize I did many things

Right: *The Terrace Standard*, Wednesday, August 29, 2007

Invention used in Olympic ads

By RICK NORTHROP

RETIRED ENGINEERING technician, Alan McGowan, is starting to get some recognition for his most famous invention.

To celebrate the upcoming 2010 Vancouver Olympics, the Royal Canadian Mint has commissioned commemorative sets of 15 coins. McGowan's invention a curling aid known as a the Curlstick, is featured on a coin with a female wheelchair curler.

"It gave me a great deal of confidence that it is a respected way of curling," said McGowan, 77, of the first time he saw the coins.

Each 25 cent coin released as part of the set portrays a different athlete and a different sport. The coins are also highlighted in a series of ads running on television and in Macleans magazine.

The Curlstick looks like a broomhandle with a hollow tube attachment that slips over the handle of curling rocks. Making use of the long handle, old, disabled or injured curlers no longer have to crouch down to cast their throws which eases strain on many an injured back or knee.

The wheelchair curler could be holding could be of any one of many Curlsticks now on the market, not necessarily McGowan's invention. Still, he is pleased his concept has become accepted.

Curlstick's invention has extended careers

BOB WEEKS
CURLING

On a list of names of those who have brought change to the game of curling in recent times, you might find Ed Werenich for his strategy, or Warren Hansen for making the Nokia Brier into a major sporting event. There would even be Lino Di Iorio, the man behind such technical advancements as the Balance Plus Slider.

And now it would be safe to add the name of Alan McGowan.

Who's that, you say?

McGowan may not win any championships and his work may not even lead to championships, but it is likely that he has brought more people back to the curling ice than anyone else in recent times.

McGowan is the man credited with inventing the Curlstick, a simple device that has allowed those with bad backs and bad knees to continue curling.

The Curlstick is similar to a shuffleboard cue: it is a long handle with a grip at one end and an attachment at the other which affixes to the curling rock. Players can use the device to deliver the stone down the ice without bending over or having to slide from the hack.

"I came up with the idea out of necessity," McGowan says. "It wasn't for me, but for all my buddies who were sitting behind the glass."

McGowan didn't start curling until he retired after 36 years of working for Alcan in Terrace, B.C. Once he stopped punching the clock, some friends convinced him to take up curling. He loved the game and the fellowship that came along with it, but over the next two or three years, all his pals were sitting behind the glass watching, unable to compete because of various ailments.

"They had bad backs and bad knees," he said. "It wasn't so much that they couldn't get down in the hack, it was that they couldn't get back up."

So McGowan, who had always been somewhat of an inventor while working for Alcan, began to tool around in his work shop. He came up with the first generation of the Curlstick and tentatively showed it to his friends.

The initial reaction wasn't that good. But after some revisions and some persuasion, it caught on.

Players who had been away from the game came back. Many who had limiting injuries went from the warm side of the glass to the cold side. Eventually, the popularity of the Curlstick grew and it spread across the country.

The device became so popular that the Canadian Curling Association made a rule change to accommodate its use. That was certainly a testament to the product's popularity, but an even greater endorsement came in 1999, at the Brier in Winnipeg, when Northern Ontario's fifth player, Paul Sauve, used the Curlstick in a game at the national championship.

Since then, many top-ranked curlers — most of whom were ailing in one way or another — have put the Curlstick into play. Or should that be a Curlstick-like device?

Although McGowan is credited with being the first to come up with the idea, many knock-offs have hit the market.

"If I was a multimillionaire, I'd probably launch lawsuits on all of them," McGowan said. "But I started it to help people. I'm not overly concerned about making money."

McGowan stated that he's sold about 500 Curlsticks over the past three years and financially the product hasn't exactly been a success. However, he still takes pride in knowing that many people who loved the game but weren't able to continue playing have returned thanks to his invention.

"It is nice to know that the Curlstick has allowed so many people to play again," he said. "That's really why I built it to begin with."

Even though it has met with approval on most fronts, McGowan still has one more aspiration.

"I'd love to get someone like [popular curler] Guy Hemmings using it," he stated. "And my real desire is to get Ed Werenich [who retired last year] back on the ice using one."

That may happen one day. For the moment, McGowan will take pride in the Curlstick being the regular guy's delivery aid.

bobweeks@sympatico.ca

Note; Warren Hansen has been the President Of the World Curling Federation and also the President of the Canadian Curling Association. Ed Werenich"The Wrench"was one of Canada's most oustanding curlers.

"outside the box," though at the time I didn't think of them that way. I just thought they would help people, and maybe I'd make a fortune. I didn't make a fortune, and my basement is full of inventions that didn't take off, but I relished all the challenges and thoroughly enjoyed myself.

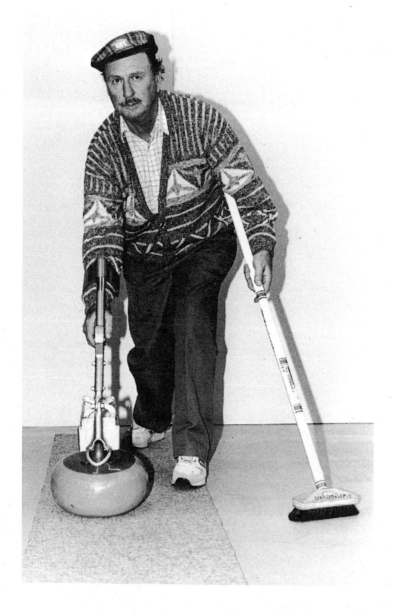

Alan on motorbike aged 15.

Epilogue

By Sharon McGowan

Just a few days after completing the final draft of this book of memoirs, my father, Alan, passed away peacefully in Terrace at age 89.

After Alan died, we searched his computer, certain that he would have written his own eulogy or obituary. He loved writing and often stepped up to create beautiful tributes to his friends and family members who had passed. We were incredulous when we didn't find anything. But then we realized he had been too busy living and enjoying the adventure, right up until the end, to ever think about himself in the past tense.

Here is the obituary (slightly edited) that I wrote for him and published in the *Terrace Standard* and *Kitimat Sentinel*:

A long-time resident of the Northwest, Alan nearly missed out on what turned out to be a long, wonderful and adventurous life. His family lived in a houseboat in Port Alice, B.C., and when he was just a few weeks old, Alan's bassinet fell off a windowsill where it was perched and into the ocean. His father, Joe, dove into the water fully clothed, emerging seconds later with a startled but otherwise unharmed baby Alan.

Growing up poor during the Depression, Alan moved with his family from Vancouver Island to Vancouver and then to Princeton, attending twelve different schools in as many years. When he was ten his parents separated, and he taught himself to raise chickens and hunt for grouse and deer to feed the family. He started working as a farm hand at 14 and by 15 had made enough money to help his family and to buy an old Harley-Davidson motorcycle for a solo road trip through California.

After finishing school, Alan worked all over B.C. as a ditch digger, miner and logger, and eventually qualified as a journeyman painter. At 24 he decided it was time to settle down. He married and with his bride, Mary, moved to the new city of Kitimat in 1954 to work for Alcan. They raised three children, Joe, Sharon and Skye, and Alan worked for Alcan for 36 years, ultimately becoming a respected project planner of large-scale innovative and ground-breaking engineering projects. When the family moved to Terrace in 1971, Alan founded the Early Riser Cooperative Bus Line between Terrace and Alcan and became well known for his daily radio reports on Highway 37 road conditions.

After retirement, Alan launched a whole new career as an inventor. His greatest success was the Curlstick, a device used all over the world now that enables curlers to play the sport while standing up. A gifted storyteller, Alan also began writing, and published his first book of memoirs, *Riding in Style, The First Twenty- Five years*, in 2015. As described by Daniel Francis, Historian and Editor of *The Encyclopedia of British Columbia*, his book is "History with a sense of humour from the perspective of someone who has lived it." After retirement, Alan also pursued his passion for inventing, and his hobby of restoring old

Alan with Mary Jane.

vehicles, completing work on a '38 Ford truck just before he died. He skied until he was 80 and rode his beloved motorcycle until he was 87.

He leaves behind a rich legacy of stories, inventions, friendships and a loving family. Alan was predeceased by his wife, Mary, in 2001, and son, Skye, in 2010, and is survived by his partner Mary Jane Hogg, ("the light of my life") and her family, his son Joe and wife Peggy, his daughter Sharon and her partner, John, and his grandchildren, Drew, Paige, Lachlan, Madeleine and Elise.